Artificial Intelligence Models for the Dark Universe

The dark universe contains matter and energy unidentifiable with current physical models, accounting for 95% of all the matter and energetic equivalent in the universe. The enormous surplus brings up daunting enigmas, such as the cosmological constant problem and the apparent distortions in the dynamics of deep space, and so coming to grips with the invisible universe has become a scientific imperative.

This book addresses this need, reckoning that no cogent physical model of the dark universe can be implemented without first addressing the metaphysical hurdles along the way. The foremost problem is identifying the topology of the universe which, as argued in the book, is highly relevant to unveil the secrets of the dark universe.

Artificial intelligence (AI) is a valuable tool in this effort since it can reconcile conflicting data from deep space with the extant laws of physics by building models to decipher the dark universe. This book explores the applications of AI and how it can be used to embark on a metaphysical quest to identify the topology of the universe as a prerequisite to implement a physical model of the dark sector that enables a meaningful extrapolation into the visible sector.

The book is intended for a broad readership, but a background in college-level physics and computer science is essential. The book will be a valuable guide for graduate students as well as researchers in physics, astrophysics, and computer science focusing on AI applications to elucidate the nature of the dark universe.

Key Features:

- Provides readers with an intellectual toolbox to understand physical arguments on dark matter and energy.

- Up to date with the latest cutting-edge research.

- Authored by an expert on artificial intelligence and mathematical physics.

Ariel Fernández (born Ariel Fernández Stigliano, April 8, 1957) is an Argentine-American physical chemist and mathematician. He obtained a Ph.D. in Chemical Physics from Yale University in record time and held the Karl F. Hasselmann Endowed Chair Professorship in Engineering at Rice University until his retirement. He was also an adjunct professor of computer science at the University of Chicago. To date, he has published over 500 scientific papers in professional journals and has also authored nine books on physical chemistry, molecular medicine, artificial intelligence, mathematical cosmology, and mathematical physics. Additionally, he holds several patents on technological innovation. Fernández is a member of the National Research Council of Argentina (CONICET) and, since 2018, heads the Daruma Institute for Applied Intelligence, the research arm of AF Innovation, a consultancy based in Argentina and the United States.

Artificial Intelligence Models for the Dark Universe
Forays in Mathematical Cosmology

Ariel Fernández

CRC Press
Taylor & Francis Group
Boca Raton London New York

CRC Press is an imprint of the
Taylor & Francis Group, an **informa** business

Designed cover image: shutterstock

First edition published
by CRC Press
2385 NW Executive Center Drive, Suite 320, Boca Raton FL 33431

and by CRC Press
4 Park Square, Milton Park, Abingdon, Oxon, OX14 4RN

CRC Press is an imprint of Taylor & Francis Group, LLC

ISBN: 978-1-032-81823-8 (hbk)
ISBN: 978-1-032-81718-7 (pbk)
ISBN: 978-1-003-50153-4 (ebk)

DOI: 10.1201/9781003501534

Typeset in Minion
by SPi Technologies India Pvt Ltd (Straive)

In loving memory of Haydée Stigliano, my mother.

Contents

Introduction

To state that the Standard Model of particle physics is successful may be an understatement. The accuracy of its predictions is simply astounding. Yet, for all its prowess, the SM can be said to have missed the dark universe as there are no hints leading to dark matter or dark energy, no obvious way to incorporate the dark sector. One can safely say that a staggering 95% of matter in all forms remains unyielding to its predictive and modelling powers.

Aided by artificial intelligence, this book addresses this conundrum. It does so by focusing on a key parameter underlying the universe that weighs pivotally on the development of the SM: The vacuum expectation value (vev) acquired by the Higgs field. Where is the vev value coming from? The answer offered by this book is somewhat perplexing: In order to break internal symmetry, the Higgs field encodes a nonzero vev in a dark (unobservable) dimension by mixing with a dark scalar field. This dark dimension curls up with a diameter corresponding to the smallest material length, assimilated to the range of the weak interaction ($\sim 10^{\wedge}(-18)$ m). Therefore, a stationary de Broglie wave spanning the dark dimension stores exactly the energy of the Higgs vev, or, alternatively, the expectation value of the vacuum state of minimum energy. Thus, the dark and visible sectors mix, generating a scalar field that induces the spontaneous symmetry breaking required by the electroweak model. This result is pregnant with possibilities, as it shows that the dark sector implicitly bestows mass to the observable universe.

In this way, the modeling effort described in the book shows that the geometric dilution of the so-called ur-Higgs, the stationary wave along the dark dimension that begets the Higgs vev, is equivalent to the Higgs mechanism within the Weinberg-Salam electroweak scheme. In general, the geometric dilution must be conserved for each particle, and as a result, additional "dark" gauge symmetries arise, superimposed onto the extant gauge

DOI: 10.1201/9781003501534-1

symmetries of the SM. Such symmetries give rise to dark bosons that communicate the force required to maintain the geometric dilution of the dark matter-wave into visible dimensions. Thus, the dark universe may be provisionally modelled through the extension of the SM that generates an $SU(2)_D \times SU(3) \times SU(2) \times U(1)$ overarching gauge theory that incorporates the mixing of the dark and visible sectors.

This effort is by no means complete and the model presented in this book should be regarded as provisional, or rather transitional, akin to Bohr's model of the atom as precursor to quantum mechanics. It is even possible, as some AI models confirm, that hitherto unknown internal symmetries of the quantum fields can account for the dark sector without the need to invoke an extra spatial dimension. A detailed and rigorous gauge theory for the dark universe will be forthcoming in a follow-up to this title. Actually, the seeding idea for such a project is the *leit motiv* for this book: As a dark compact dimension is incorporated, the relativistic wave equation for dark-sector fermions casts the mass terms as kinetic-energy operators composed so as to represent geometric dilution into the visible dimensions when applied to Standard Model particles.

Ariel Fernández
Riverside, Connecticut, USA

CHAPTER 1

The Dark Universe

Metamodel and Foundational Evidence

"*Φύσις κρύπτεσθαι φιλεῖ.*"
HERACLITUS

(*Nature likes to keep its secrets*,
TRANSLATION BY THE AUTHOR)

This chapter provides a metamodel of the dark universe; that is, of the matter and energy unidentified in current physical models and assumed to be present in overwhelming amounts in deep space. By metamodel we mean an inferential scheme that incorporates big evidentiary data and generates a matrix for a cogent and physically consistent model capable of meaningful prediction. The evidence includes large-scale phenomenology that supports the existence of invisible matter and energy, established to contribute significantly to steer the dynamics of deep space. Their overpowering abundance and lack of interaction with gauge bosons that communicate quantum forces are likely to trigger a paradigm shift in physics.

Leveraging fairly elementary physics arguments, we examine experimental data on star rotation in galaxies, where anomalously high rotational speeds lead us to conjecture the existence of dark matter. We also describe evidence supporting the accelerated expansion of the universe, pointing to the existence of dark energy.

DOI: 10.1201/9781003501534-2

The chapter also surveys the Dirac equation under the assumption that it provides a vantage point to interpret the dark universe, as demonstrated in Chapter 5.

1.1 INTRODUCTION

This book focuses on the dark universe, that is, on the vast proportion of matter and energy in deep space that cannot be accounted for by extant physical models. To shed light on the problem, artificial intelligence (AI) is leveraged, and testable assumptions are made concerning the nature of dark matter and dark energy. As described in the subsequent chapters, cosmological data will be inputted into an appropriate AI system under the assumption that there exists a "dark dimension" such that any particle field concentrated along this dimension will not interact with photons unless there is a spillover or "geometric dilution" of the field into the observable dimensions.

This picture prompts us to tentatively model the universe adopting a generic AI system under the guise of what is known as an *autoencoder*. In the standard and rather restricted view, autoencoders can be thought of as embodiments of machine learning systems suited to predict dynamical behavior of a physical system from a time series of a set of observables. This time series can be spliced into a training set and a testing set. As described in Chapter 3, the predictive power of autoencoders hinges on distilling the dynamics into a simplified version named "latent dynamics" which is assumed to be operative in a lower dimensional space (usually a differentiable manifold), known as "latent space." The dimensionality reduction is presumed to capture the essential dynamics; that is, the modes that enslave fast-relaxing degrees of freedom that can be averaged out for predictive purposes.

To illustrate the concept of autoencoder, suppose the universe were a giant molecule, an assemblage of N atomic nuclei and J electrons. In a classical physics view, the state of the system at time t would correspond to a point $x = x(t)$ in the space $W = \mathbb{R}^{3(N+J)}$ and a time series training the AI system would be given as $\{x(n\tau), x((n + 1)\tau)\}_{n = 0, 1, ..., M}$, where the period τ indicates the level of time coarse graining. Then, the prediction would entail an estimation of $x(L\tau)$, $L = M + 2, M + 3, ...$ By making physically reasonable assumptions, it is possible to simplify the dynamics with significant dimensionality reduction, thereby enabling the prediction. For example, it is well known that the electron "motion" can be treated "adiabatically," that is, in an averaged way relative to the nuclei. This is so since the electron degrees of freedom are entrained by those of the nuclei. This is the so-called Born-Oppenheimer (B-O) approximation adopted in molecular and atomic physics, as described in Chapter 3, and enables the autoencoder to adopt a latent manifold $\Omega = \mathbb{R}^{3N}$, thus significantly reducing dimensionality by averaging out the $3J$ electron degrees of freedom.

A more generic way of visualizing the encoding, more akin to particle physics, would be to consider the level of delocalization (uncertainty in the determination of position) of the particles in the assemblage. The delocalization is given by the so-called Compton wavelength defined as $\lambda_C = \dfrac{1}{2}\dfrac{h}{mc}$, where h is Planck's constant, c is the speed of light, and m is the particle mass. Thus, it becomes clear that, relative to the nuclei, the electrons may be represented as "clouds" with averaged-out motion, whereas the nuclei, with their far larger mass and hence significantly smaller Compton wavelengths, may be treated as localized objects.

Based on the B-O *ansatz*, we may construct an autoencoder adopting a multilayered neural network technology (Chapter 3), so that the dynamics in W becomes encoded or simplified at a coarser but fundamental level in Ω and subsequently decoded back to W (Figure 1.1). Essentially, if $F : W \rightarrow W$ represents the time evolution map such that $F\boldsymbol{x}(n\tau) = \boldsymbol{x}((n + 1)\tau)$, the autoencoder is determined by a tern (γ, K, μ) consisting of the map $K : \Omega \rightarrow \Omega$, the projection $\gamma : W \rightarrow \Omega$, and the hologram (injection) $\mu : \Omega \rightarrow W$, optimized in

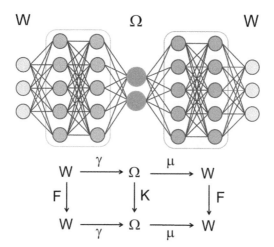

FIGURE 1.1 Observable universe (Ω) modelled within an AI system under the guise of an autoencoder constructed using multilayered neural network technology and defined by the tern (γ, K, μ). The dark universe becomes the portion of a higher-dimensional universe W that incorporates a dark dimension and excludes the hologram $\mu\Omega$. The autoencoder parameters are optimized so that the diagram at the bottom becomes commutative.

such a way that the following commutativity relations hold: $K \circ \gamma = \gamma \circ F$ and $F \circ \mu = \mu \circ K$ (Figure 1.1). In other words, the autoencoder is able to predict the output $x((L + 1)\tau)$ from input $x(L\tau)$ as $x((L + 1)\tau) = (\mu \circ K \circ \gamma)\, x(L\tau)$.

Human civilization has been able to physically model the fundamental particle fields that constitute the fabric of the universe, as described by the so-called Standard Model (SM) of particle physics. However, the model clearly does not encompass the dark portion of the universe, the vast portion now known to be impervious to interaction with photons and other gauge bosons in the SM. Thus, we may regard the extant physical model as actually corresponding to the evolution map on a latent manifold $K_{SM} : \Omega \to \Omega$, but not on W. Un-localizable (undetectable) particles, altogether lacking Compton wavelength, are excluded from the latent physics, requiring (AI postulates) a dark dimension to materialize.

We know that the K_{SM} model is unsatisfactory in deep space because it does not incorporate the bearing of the dark universe on the detectable dynamics, as described in this chapter. When this influence is incorporated, especially at the large scales of deep space, we end up with a model of broader applicability $K_{SM}^{\odot} : \Omega \to \Omega$ with $K_{SM} \neq K_{SM}^{\odot}$. Under the universe-as-autoencoder *ansatz*, determining the physics of the dark universe becomes tantamount to identifying the dark universe as the large portion, $W \backslash \mu \Omega$ of W that excludes the hologram $\mu \Omega$ and determining the autoencoder-model (γ, F, μ) that satisfies the commutative relations $K_{SM}^{\odot} \circ \gamma = \gamma \circ F$ and $F \circ \mu = \mu \circ K_{SM}^{\odot}$. This is the overarching goal of the book.

1.2 METAMODEL OF THE DARK UNIVERSE

Twentieth century physics witnessed two major advances: Relativity and quantum mechanics, both representing in their own right towering achievements in the understanding of the universe at what appeared to be vastly different scales, from cosmological to subatomic. Yet, as mankind came to grips with cosmological singularities such as black holes, the locality and dimensions under consideration enabled the possibility of a common ground for quantum gravity, purportedly bridging the two towering theories [1]. At the same time, the awareness that a vast proportion (95%) of the universe contains invisible or undetectable forms of matter and energy added further impetus to the pursuit of quantum gravity as a vantage point to understand unyielding aspects of deep space and interpret the phenomenology that arises thereof. The hitherto unknown forms of matter and energy constitute the dark universe and are likely to trigger a paradigm shift that may shake the edifice of physics at its very foundations [2, 3]. This assertion served in good measure as the motivation for this book.

Three pivotal tenets, justified and elaborated in the subsequent chapters, define our approach to the dark universe:

I. The topology of space-time is a determinant of the nature of the dark universe.

II. The universe is compact and cannot have boundaries because there is no interface with nothingness, therefore space-time is a multiply connected manifold.

III. The universe contains a hologram in the sense that the quantum fabric of space is apparent within a dimension reduction yielding a latent manifold that encodes the relativistic space-time [4].

The last tenet implies that an AI–system in the guise of an autoencoder may be ideally suited to study quantum gravity [5]. Autoencoders may be thought of as learning systems that distill the fundamental physics underlying a process or phenomenology by encoding it in a lower dimensional manifold, known as latent manifold. Thus, *a hologram becomes styled as an autoencoder* in the AI context. A physical embodiment of the autoencoder may be tuned with adherence to statistical mechanical laws in such a way as to "realize" quantum gravity [5].

The sheer existence of a dark universe made up of invisible forms of matter, including dark energy, summons a hologram setting for the universe, in which the phenomenology associated with the dark universe manifests itself in the latent manifold (Ω) only indirectly and through gravitational perturbation. This manifold corresponds to the realm of the detectable, the realm where, in terms of quantum physics, an act of observation collapses the wave function. However, as shown in this book, the dark universe is contained in a higher dimensional space (W), an ur-universe, that admits no observer. Only 5% of matter in all forms in W is encoded by Ω, with the remaining 95% corresponding to the dark universe. We may state that "*the dark universe is the part of W that is not part of the hologram.*" In rigorous terms, if $\eta : \Omega \to W$ is the holographic function, the dark universe $\mathcal{W}$ is defined as $\mathcal{W} = W \setminus \eta(\Omega)$.

Let $K_{SM}^{\circlearrowleft} = \mathcal{K}_{SM} : \Omega \to \Omega$ define a generic transformation in the Standard Model of particle physics, then the overarching goal of this book is to identify the transformation $\mathcal{F} : W \setminus \mathcal{W} \to W \setminus \mathcal{W}$ and its extension to W, jointly with the map $\gamma : W \setminus \mathcal{W} \to \Omega$ such that the diagram in Figure 1.2 is commutative, which means that $\mathcal{F} = \eta \circ \mathcal{K}_{SM} \circ \gamma$.

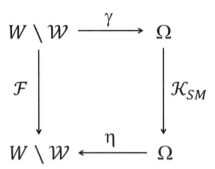

FIGURE 1.2   Metamodel of the dark universe $\mathcal{W} = W \setminus \eta(\Omega)$ understood as an AI system in the form of an autoencoder. The dark universe is identified with the portion of the ur-universe W that excludes the hologram, $\eta(\Omega) \subset W$, encoded by the latent manifold Ω. The dark universe is characterized by the lifting $(\mathcal{F}, \gamma)$ of a "corrected" Standard Model map $(\mathcal{K}_{SM}, \eta)$ that makes the diagram commutative in the sense that $\mathcal{F} = \eta \circ \mathcal{K}_{SM} \circ \gamma$.

The dark universe is thus determined by the compatibility of the observable universe (Ω, η), and the ur-universe (W) guaranteed by the commutativity of the diagram in Figure 1.2, which represents the metamodel for the dark universe.

This picture is incomplete and, in a sense, unsatisfactory insofar as the ur-universe remains undefined. To grasp the nature of the problem we need to incorporate tenets I and II. The topology of the universe prompts the autoencoder to determine the embedding of the 4-dimensional space-time W_4 into the higher dimensional space W. As discussed in the subsequent chapters, the spatial dimensions in W_4 must be compact because they once were, if we grant validity to the big bang scenario in a relativistic context, and relativistic space-time cannot afford topological change. Furthermore, Ω cannot have a boundary since that would imply interfacing with nothingness, that is, not with a vacuum but with absence of geometry. Such an interface represents a physical and metaphysical impossibility. As a compact manifold without boundaries, Ω must be multiply connected and, consequently, so must be its hologram $\eta_4(\Omega) \subset W_4$. However, as shown in Chapter 5, to retain rotational symmetry (more specifically, axis permutation), the multiply connected space-time must be endowed with a primeval wormhole. This singularity, however, should be regarded as artifactual of the dimensionality of the space that admits the quantum observer. The situation is akin to that of the Klein bottle, a non-orientable surface that features self-intersection when displayed in the 3-dimensional space but not in four dimensions, where it is not amenable of representation (Figure 1.3a). Furthermore, as shown in Chapter 4, W_4

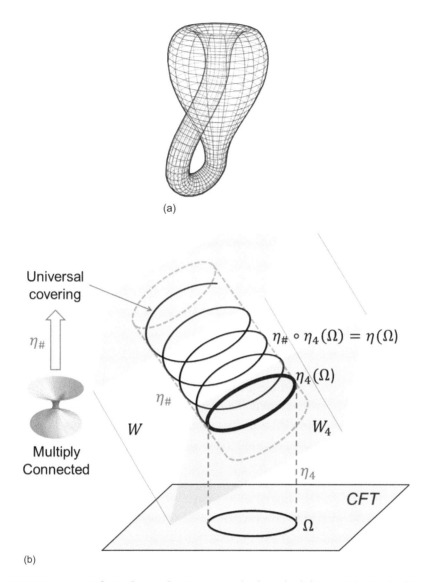

(a)

(b)

FIGURE 1.3 Artifactual singularities smoothed out by lifting at the level of the universal covering. (a) The Klein bottle, a non-orientable 2-dimensional manifold that self-intersects when represented in a 3-dimensional space, but not in four dimensions. (b) The multiverse or ur-universe W topologically represented as the universal covering of a holographic autoencoder.

should be embeddable in a space, W_5, with an extra spatial compact dimension required to store dark matter in the form of stationary waves with no geometric dilution into the other spatial dimensions.

The ur-universe W becomes then the universal (simply connected) covering of W_5. In this way, the simply connected topology that smooths out

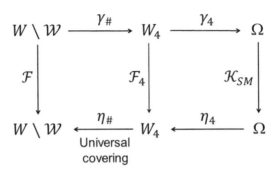

FIGURE 1.4   Characterization of the dark universe through a cogent metamodel of the universe represented as an autoencoder. The compatibility between the holographic and lifted (multiverse) representation of the latent physics enshrined in the Standard Model is warranted by the commutativity of the diagram.

the inherent singularities of W_5 is recovered at the expense of incorporating a noncompact spatial dimension (Figure 1.3b). This implies that the holographic map $\eta : \Omega \to W$ can be factorized as $\eta = \eta_\# \circ \eta_4$, through the embedding of the latent space $\eta_4 : \Omega \to W_4$ composed with the lifting $\eta_\# : W_4 \to W$, which is a multivalued map. This implies that *the universal covering W (together with the lifting $\eta_\#$) corresponds to an AI–encoding of the many-worlds interpretation of quantum mechanics*. This interpretation is thus validated by AI through the commutativity of the diagram presented in Figure 1.4, whereby the dark universe is identified as $\mathcal{W} = W \setminus (\eta_\# \circ \eta_4)(\Omega)$ and the following relations hold: $\mathcal{F} = \eta_\# \circ \mathcal{F}_4 \circ \gamma_\#$, $\mathcal{F}_4 = \eta_4 \circ \mathcal{K}_{SM} \circ \gamma_4$, where $\gamma_\#$, γ_4 are, respectively, the projections associated with the lifting and the holographic map.

One would say that the overarching goal of the book is to determine the maps $\mathcal{F}, \gamma = \gamma_\# \circ \gamma_4$, that make the diagram in Figure 1.4 commutative and thereby enable the elucidation of the dark universe.

1.3 THE DARK UNIVERSE: A NAÏVE PERSPECTIVE

Radiation and matter have been regarded as interrelated concepts since humans pondered nature and the cosmos, way before the advent of physics as a formal discipline. One of the latest musings over this relationship came with Albert Einstein, who enquired how could the photon, the quantum of light, carry momentum while it has no mass. That perplexity eventually led him to formulate what may well be the most famous equation in all of physics: $E = mc^2$, with $E =$ energy, $m =$ mass at rest, $c =$ speed of light. The idea of finding the massive particle equivalent to the photon also led him to the idea that gravity may affect radiation, which had no clear precedence in physics.

Rather than writing the momentum of the photon in standard form as $p = mc$ (p = momentum), he chose the more fundamental expression $p = E/c$, where E is the "kinetic energy" of the photon, which, as shown by the physicist Max Planck (1858–1947), is proportional to its frequency f as radiation carrier. The proportionality factor being Planck's constant h, so that we may write $E = hf$. But the crux of Einstein's argument is the question: What would be the mass of a particle that would carry the same momentum as the photon? Reciprocally, we may ask what would be the frequency of the photon "equivalent" (i.e., with the same momentum) to a particle of rest mass m. Based on Einstein's relation – and assuming momentum to be far smaller than mc – the answer at hand is $f = \dfrac{mc^2}{h}$.

Now, what if we suddenly decide to postulate that only 5% of matter, made up of particles endowed with mass or light quanta equivalents, actually obeys the relation $f = \dfrac{mc^2}{h}$? That would be indeed a shocking statement meeting with immediate skepticism and yet it reflects the current state of affairs in the realm of deep space cosmology [1]. We shall provisionally name matter that does not interact, exchange, or communicate with radiation "dark matter (DM)," and reciprocally, we shall term "dark energy" (DE) the energy that cannot find a material equivalence.

It is clear that the Einstein relation did not hold in a big bang scenario for the origin of the universe: If it did, radiation would have evened out the distribution of mass in the cosmos and would have smoothened out fluctuations in the radiation relic of the big bang detectable today and known as "cosmic background radiation" (CMB) [1]. So, we may conjecture that dark matter and dark energy originated in an early universe and prevailed, so that Einstein's relation is upheld only for 5% of matter, for what we have aptly chosen to name "detectable" or "visible" matter. This book is devoted to elucidating the nature of the vast proportion of matter in the cosmos that does not communicate or interact with light quanta, the "dark cosmos."

Physics is an established and respected field of knowledge that endeavors to explain how the universe works at all scales. Its corpus incorporates ideas, models, and data only after careful scrutiny. The bar is high and scientists who want to leave their mark face a stringent peer review process and only get to impose their views after a hard-won battle. Solid as it seems, there is nothing monolithic, no final word in physics. Truth is never fully conquered but stands as a beacon for the daring. Each time there is a breakthrough, the veil of mystery is lifted a little, but as the horizon expands, new mysteries arise. Every concept, every theory, every measurement is constantly subject to revision as new windows of reality open up to detection and technologies are endlessly perfected. Paradigms, even those that appear rock hard for a while and endure a long-term attrition, often crumble under the weight of new and disconcerting evidence or undergo extensive revision. The history of physics is endlessly made of cycles of destruction and creation, much like the ancient cosmogonies of the valley of the Indus, dictated by the infinite toils of Shiva, the destructive element that dances out the pulse of the universe.

In or around 1900, the Irish physicist William Thomson (1824–1907), usually referred to as Lord Kelvin, famously declared: *"There is nothing new to be discovered in physics now. All that remains is more and more precise measurement."* In the two or three decades that followed after this pronouncement, two earth-shattering revolutions in physics took place: Einstein's theory of relativity and quantum mechanics [1]. So much for Lord Kelvin's solemn pronouncement …. History proved once again to be the master of irony.

Relativity and quantum physics thrived and prospered because they effectively and successfully addressed shortcomings in the prevailing paradigm at the turn of the 20th century. This paradigm is essentially enshrined in two basic pillars of knowledge: a) Newton's law of gravitation that governs the dynamics of falling and orbiting bodies at terrestrial (the apocryphal falling apple) and cosmic scales and b) Maxwell's laws governing electromagnetic phenomena, that is, the events that reveal the entanglement between electricity and magnetism [1]. It thus seemed that all the effects involving the known forces in the universe at the time were satisfactorily understood, even if the nature of such forces remained unyielding to theoretical efforts. For example, when asked about the nature of gravity, Newton snapped in Latin: *"Hypotheses non fingo"* (I contrive no hypothesis). In an ironic turn, right after Lord Kelvin's pronouncement, Einstein came up with relativity, the first theory that truly explained gravity, while an avalanche of new data revealed that matter at atomic and subatomic scales required a drastic revision of the extant conceptual framework, heralding the birth of quantum mechanics.

Einstein showed that time and space are inevitably entangled and should not be treated separately, with time acting as the fourth dimension.

His view of gravity in "space-time" is admirably synthesized in the quote of the American physicist John Archibald Wheeler (1911–2008): "Space-time tells matter how to move, matter tells space-time how to curve" [1]. On the other hand, quantum mechanics successfully accounted for the discontinuous nature of energy and momentum experimentally shown to hold when the behavior of matter is observed at atomic and subatomic scales. Thus, changes in those physical magnitudes are often accompanied by emission or absorption of radiation which can only vary in a discrete fashion, as multiples of a constant. These packages of discontinuous energy are known as quanta, a term coined by the German physicist Max Planck (1858–1947).

Lofty and sturdy as it may seem, the edifice of contemporary physics is beginning to show major cracks and may not withstand the attrition to which it is now exposed. The cracks – we now know – are not superficial but extend all the way to the very foundations. As the dynamic structure of the universe is examined at very large scales, commensurate with the dimensions of galaxies, and further, of clusters and other assemblages of galaxies, major anomalies are surfacing. In fact, the anomalies have been surfacing since the 1930s. Now we know the cracks are indeed structural and require immediate attention or the whole edifice of physics, the crowning achievement of human civilization, may be doomed and per-haps even tagged for a paradigmatic demolition.

Numbers simply don't add up in the cosmos. In the outer shells, stars in spiral galaxies have been behaving in ways that can be considered anomalous – to use physicists' typical euphemism – spinning at speeds far larger than those that would enable gravity from the visible universe to hold them in stable orbits. From this perspective, for the universe to make sense, the gravitational pull must be far larger than what is expected from the amount of matter detected [1, 2]. This begs the question: Where is the missing matter? Or are we supposed to introduce a fudge term in the equa-tions? This disquieting picture of galaxy rotation became apparent through the pioneering work of the American astronomer Vera Rubin (1928–2016). Vera examined six spiral galaxies, like our Milky Way, or nearby Andromeda, and consistently found outer stars behaving anomalously: There is simply not enough detectable mass to keep them in their orbits. Newton's law of universal gravitation worked astonishingly well when it comes to describe the dynamics of our solar system – which earned Newton a superlative reputation – but when it comes to stars revolving around the center of galaxies, we are dismayed by the outcomes of the theory. We are left with two painful alternatives: Either the law of universal gravitation fails miserably or there is a huge amount of matter out there that we cannot detect and introduces a considerable gravitational pull, five times larger than ordinary matter [1–3].

This is not the only problem with the cosmos – not even the worst one – that physicists need to cope with: The universe is expanding at an ever-increasing rate that far exceeds what the gravitational pull would enable and far exceeds the amount of kinetic energy available from the motion of visible matter. It seems that plenty of the matter required to sustain the structure and dynamics of the universe in accord with Newtonian or even relativistic laws simply cannot be accounted for. It is simply invisible. The missing matter has been named "dunkle Materie" (dark matter) by the Swiss astronomer Fritz Zwicky (1898–1974) who first postulated its existence, although perhaps "invisible matter" would have been a better name [3].

More troublesome, the rate of expansion of the universe is not constant: It is going exponential (Figure 1.5). As their speed exceeds the speed of

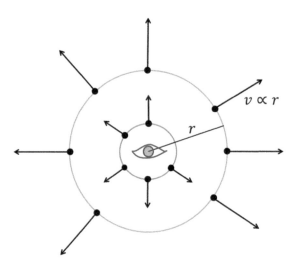

FIGURE 1.5 Schematic representation of a universe undergoing accelerated expansion. The component of the velocity of a celestial object contributing to its getting away from the observer increases as the object is farther away from the observer. Distant objects (larger radius r) have been traveling for longer times and hence their speed has had a chance to increase more relative to those that have travelled less.

light, zillions of stars will simply spin out of our horizon and become forever undetectable: Their light will never reach us. For this staggering discovery, astrophysicists Saul Perlmutter, Brian Schmidt, and Adam Riess were awarded the 2011 Nobel prize in physics [1, 3]. What is the source of this enormous surplus in kinetic energy seemingly sprouting from the vacuum of deep space? Nobody is certain, although a quantum mechanical origin is often invoked, hence the term "quantum vacuum fluctuations." An investigation on the nature of this mysterious energy gets too technical and is deferred to Chapter 5. This view postulating quantum vacuum energy is problematic, to say the least, since the naïve energy estimations are off by a factor of 10^{120} when contrasted against the experimental results obtained by Perlmutter, Schmidt, and Riess! The surplus kinetic energy associated with the berserk expansion of the universe has been named "dark energy."

Dark matter and dark energy cannot be considered corrections to the laws that presumably govern the cosmos [2, 3]. There is roughly five times more matter in the universe than visible matter, the matter we are able to detect. We know this from the gravitational influence of dark matter on visible matter. Furthermore, we know that energy and matter are interchangeable, with one becoming a proxy for the other and vice versa, as implied by the arch-famous Einstein equation $E = mc^2$ mentioned previously in this section. With this formula in mind, the breakdown of the actual mass budget of the universe looks even weirder. Give and take a few tens of a percent, current state-of-the-art calculations yield the following composition of the universe:

Ordinary matter: 5.0%

Dark matter: 26.7%

Dark energy: 68.3%

These ratios are nothing short of scandalous. We need to come to grips with the fact that the label dark matter simply captures our ignorance regarding the nature of most of the matter in the universe. Shockingly, 95% of the universe may be accounted for but remains utterly undetectable. We infer its existence indirectly, through its gravitational influence on ordinary matter.

Invisible matter, now present at temperatures almost negligible (3 K, or −454 F, or −270 C), originated at much higher temperatures prior to the formation of galaxies, yielding the name "cold dark matter" from the fact that it moves at nonrelativistic speeds ($v \ll c$) [2]. This material of unknown nature, first accrued into small galaxies and subsequently served as building-block and seeding or nucleating material for larger scale structures up to the present-day gravitationally bound clusters of galaxies. In the widely accepted standard cosmology model, the gravitational growth of present-day galaxies and their clustering is steered by primeval fluctuations that animated a sea of cold dark matter. Although the nature of dark matter is anyone's guess at this point, astrophysicists have measured the imprint of their fluctuations in their primeval spatial distribution.

Such ancient imprint is embossed as slight variations across the universe in the brightness of the so-called cosmic microwave background (CMB, mentioned earlier), the relic ultra-weak radiation field (Figure 1.6) left over from the big bang [1]. The CMB is the landmark evidence of the big bang origin of the universe and constitutes a faint radiation that fills up all space, dating back to the time when the atoms were first formed.

FIGURE 1.6 Cosmic Microwave Background (CMB) displaying variations in the intensity of the radiation. The CMB is a snapshot of the oldest light in our universe, imprinted on the sky when the Universe was a mere 380,000 years old. The signal is faint and lies in the radio wave spectrum. It shows tiny temperature fluctuations that correspond to regions of slightly different densities, representing the seeds of all future structure in the universe: The stars and galaxies of today. Credit: US National Aerospace Agency (NASA). Image from the public domain. (https://commons.wikimedia.org/wiki/File:Ilc_9yr_moll4096.png)

With optical telescopes, the space in the background of light-emitting objects is completely dark. Only a very sensitive radio telescope shows a faint background noise, a glow not associated with any object in the sky, a signal that is strongest in the microwave region of the spectrum. Its accidental discovery in 1965 is credited to American astronomers Arno Penzias (1933–) and Robert Wilson (1936–), earning them the 1978 Nobel prize in physics [1]. In colloquial terms, we may say that we are still "hearing" that massive explosion as the CMB (the radiation is in the radio wave frequency range). Spatial changes in the intensity of what we are hearing may give us clues on how dark matter got organized and distributed after the big cool down that followed the big bang. The acoustic oscillations detected experimentally to exquisite precision in the brightness fluctuation smudges of the CMB indicate the presence of a dominant invisible form of matter that flows freely, noble and inert, alongside the ordinary matter and radiation that are tightly coupled through electromagnetic interactions.

Today many experiments are frantically searching for signatures of dark matter, both in the sky and in the laboratory, including the widely known Large Hadron Collider (LHC), a massive international consortium built near Geneva, Switzerland, to discover and detect subatomic particles [2]. This search has so far been unsuccessful and this book does not conceal some skepticism regarding the outcome of LHC experiments in regards to dark matter, as discussed in subsequent chapters.

As suggested by the path-breaking work of Vera Rubin, the revolution dynamics of stars and dust in galaxies imply the existence of invisible mass in a halo that extends well outside the inner region where ordinary matter concentrates [1, 3]. Surprisingly, the need for dark matter in galaxies appears only in the outer region where the gravitational acceleration drops below a universal value, which equals roughly the speed of light (299,792.458 kilometers or roughly 186,000 miles per second) divided by the age of the universe (13.82 billion years = 436,117,077,000,000,000 seconds). This is a highly disconcerting fact within the favored interpretations of dark matter. The sheer existence of a universal threshold (0.0000000007 meters per second squared) in the acceleration due to gravitation raises the daunting possibility that we are not actually missing matter but rather witnessing a change in the effect of gravity on the dynamics of visible matter at extremely low values of the gravitational force. This would require a very significant revision of the cosmology models enshrined in the theories of Newton and

Einstein, a revision that may be tantamount to a dramatic paradigm change, bringing shockwaves to the scientific community.

1.4 LEVERAGING AI TO EXPLORE THE DARK UNIVERSE

More definitive clues are needed to figure out the nature of dark matter and dark energy. This book squarely addresses this imperative. In their quest, humans have at their disposal and also carry the burden of an enormous corpus of knowledge across a variety of fields, from particle physics to cosmology. Thus, a solid command of the so-called Standard Model in particle physics may surely become a blessing as it provides a foundational substrate to build upon, but in some sense it may also become a handicap since the sheer volume of knowledge may hamper the boldness of approach that the dark matter/dark energy conundrum demands. Unless we get a smarter civilization to whisper the answer to us, it seems that at this point the problem is eminently suitable for artificial intelligence, which does not have to pay respects to tradition and can embark in the boldest assumptions without other restraints than those imposed by logic and consistency. This is precisely the approach adopted and described in this book in the most elementary way possible for the benefit of a broad audience.

The approach resorts to the underlying science critically revisited by artificial intelligence in the light of the startling anomalies observed experimentally. When properly steered, artificial intelligence illustrates how science is best done: Not by prejudice, but through analysis of new big data in response to an intriguing hypothesis that humans, too loyal to their scientific cliques, are sometimes unable to formulate or accept.

As argued previously, as we prod the cosmos at very large scales, basic tenets of physics seem to crumble under the weight of contradicting evidence. This book helps mitigate the current crisis. It resorts to artificial intelligence for answers and describes the outcome of this quest in terms of an ur-universe, a quintessential compact and multiply connected space that incorporates a fifth dimension to encode space-time as a latent manifold.

The American physicist John A. Wheeler aptly characterized Einstein's universe with the phrase: *"Matter tells space how to bend while space tells matter how to move."* But it turns out that at the largest cosmic scales, plenty – the vast majority – of matter and movement goes unaccounted, so either there is a colossal surplus of dark matter and dark energy that cannot be detected, or Einstein's theory becomes inadequate to explain deep space. Humans cannot judiciously decide at this juncture, so we must let artificial intelligence be the arbiter.

In some ways, AI is bolder than humans because the huge corpus of knowledge, starting with the prodigious Standard Model of particle physics, poses almost no burden to its conjecture-framing processes. So, the plan set forth for the rest of the book is to feed AI with the SM enriched with the troubling cosmological phenomenology on dark matter and dark energy and see how AI reconciles the seemingly conflicting data with the currently accepted laws of physics. This is in a nutshell the intellectual adventure that lies ahead for the reader.

1.5 DARK MATTER IN DEEP SPACE

As keen observers of the cosmos, we are now taking a closer look at dark matter. Due to its lack of interaction with light quanta, we cannot expect to see or detect dark matter directly, but we must be prepared to infer its existence from its influence on ordinary (visible) matter. This influence is known to be solely gravitational, as dark matter does not appear to be implicated with other forces of nature. Essentially, we shall strive to give physical reasons why we think that there is such a thing in the first place. Let us consider the best understood and simplest possible celestial motion, that is, that of a planet orbiting around a star. For simplicity, the motion may be assumed to be circular, with the star at the center. Thus, the planet moves in its orbit with velocity v and is located at all times at distance r from the star. For the sake of illustration, we assume the star is our sun and we are describing planetary orbits in our solar system. This picture is only an approximation, as planets typically describe elliptical orbits with the sun in one of the foci of the ellipse. A key physical quantity in this picture is the *centripetal force* – let's name it F – that is, the force toward the center of the circular motion provided by the gravitational pull of the sun and required to keep the planet in its stable circular orbit. This force was first calculated by Newton from the simple rigorously obtained relation $F = mv^2/r$, where m is the mass of the planet, v is its velocity along its circular orbit, and r is the – constant – distance between orbiting planet and sun or, equivalently, the radius of the circular orbit, whichever the reader prefers. As said, this force is provided by the gravitational pull of the sun, hence it is equal to the gravitational force, which according to Newton's law of universal gravitation is given by $F = GMm/r^2$, where M is the mass of the sun and

G is Newton's gravitational constant [1]. By combining both expressions of the centripetal force we get

$$F = \frac{mv^2}{r} = \frac{GMm}{r^2}.$$

(1.1)

Equation (1.1) yields $v^2 = GM/r$, or $v = \sqrt{GM/r}$. This result is very important to describe the solar system since it implies that the further away a planet is from the sun, the slower the speed at which it revolves around the sun [1]. We can qualitatively plot this planetary behavior for the solar system as shown in Figure 1.7.

Now let us consider a much larger scale in the cosmos. Instead of planetary systems, let us consider galaxy rotation, or more precisely, the dynamics of stars that revolve around the center of spiral galaxies. We would imagine that this motion would be similar to that of planets revolving around stars, but that is not the case: Fundamental anomalies arise, and – as it turns out – these anomalies are key to infer the existence of dark matter.

Since the visible mass M of the galaxy exerting a gravitational pull on the star is mostly concentrated at and "near" the center of the galaxy, we are likely to expect a v-r correlation similar to that presented in Figure 1.7, with v being now the speed of the star and r the distance to the center of the galaxy. In this case, our prediction would extend to "short" galactic distances where the mass contained inside the radius of the star orbit, the mass that exerts gravitational pull on the star, becomes significantly

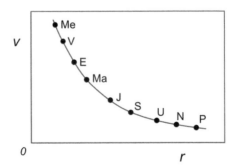

FIGURE 1.7   Qualitative scheme of the correlation between velocity and distance to the sun for the nine planets of the solar system (Me = Mercury, V = Venus, E = Earth, Ma = Mars, J = Jupiter, S = Saturn, U = Uranus, N = Neptune, and P = Pluto).

smaller as we approach the center of the galaxy as more and more mass is left outside the star orbit. Based on experimental observation of visible matter in the heavens, we are assuming that the mass is concentrated at and near the center of the galaxy and that, as we navigate toward the boundaries of the galaxy, mass concentration is negligibly small relative to the concentration near the center.

Like in planetary motion, the behavior of the revolving velocity of the star is predicted again to be $v^2 = GM/r$, except that now M decreases substantially for small r, as less visible matter lies inside the star orbit with decreasing r. Let us get specific at this point: The detailed prediction for spiral galaxy Messier 33 is presented in Figure 1.8. On the other hand, experimental work paints a very different picture: Observation of the galaxy following the pioneering work of Vera Rubin reveals a very different behavior, particularly for distant or outer stars in the galaxy, as shown in Figure 1.9. The measured velocities of the outer stars are seemingly less dependent on the distance to the center of the galaxy [1]. For outer stars, as the distance r to the center of the galaxy increases, M the mass contained within the orbit of the star hardly changes at all since, as we have said, most of the mass is concentrated at or near the center of the galaxy. Yet, the velocity of the star is no longer decreasing according to the Newtonian formula $v^2 = GM/r$ (Figure 1.9). The experimental measurements shown in Figure 1.7 have been corroborated time and again by independent observers around the world and are strikingly similar for all examined spiral galaxies [1]. The results invariably show that outer stars in galaxies are behaving anomalously: They are spinning so fast that the

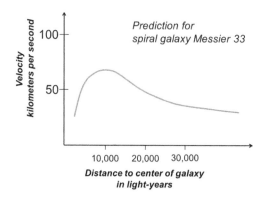

FIGURE 1.8   Predicted correlation between velocity of a star revolving around the center of the galaxy and its distance to the center of the galaxy. The curve was obtained for the spiral galaxy Messier 33.

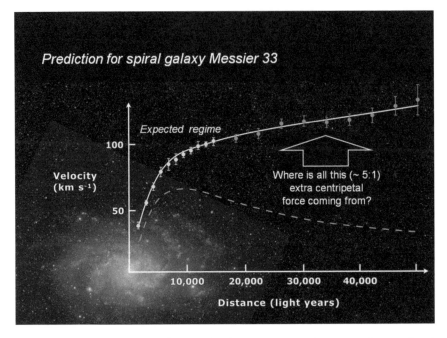

FIGURE 1.9 Rotation curve for spiral galaxy Messier 33 (points with error bars) and predicted curve from distribution of the visible matter (dashed gray line). The prediction and experimental measurement give the velocity of star rotation around the center plotted against distance of the revolving star to the center of the galaxy. The huge discrepancy between theoretical and experimental values is attributed to dark matter distributed in a halo surrounding the galaxy. Adapted from a figure in the public domain. Credit: Mario De Leo – Own work, CC BY-SA 4.0, https://commons.wikimedia.org/w/index.php?curid=74398525.

centripetal force or gravitational pull exerted by the galaxy would not be enough to keep them in orbit, yet the stars are clearly revolving around the center in stable orbits. According to Newtonian physics, at the measured speeds the outer stars in the galaxies should fly away into outer space. This is clearly *not* what is happening.

Taken together, these observations clearly and unambiguously suggest that Newtonian physics breaks down; it is no longer upheld. The only alternative that would bring some intellectual relief to the physics community is that the mass M of the galaxy, responsible for the gravitational pull on the star, has been grossly underestimated. But how could that be? All detectable matter has been detected and the mass computed with satisfactorily convergent results from constantly perfected technologies. These anomalies prompted physicists to postulate the existence of dark matter, that is, undetectable matter of an unknown nature distributed as a halo around the galaxy. This type of matter would contribute significantly to the gravitational pull necessary to ensure the stability of the orbits of outer stars.

Thus, the only inevitable conclusions from results such as those shown in Figure 1.9 are that either a) the physics that withstood centuries of attrition needs a major correction or b) the mass M in the gravitational equation that governs the motion of the stars in the galaxy represents something else, something quite different from the mass of the ordinary (detected) matter. In either case, both assumptions would represent fundamental departures from what we have been taking for granted for the last three hundred years.

We could in principle modify the sacred Newton's gravitational law $F = GMm/r^2$ for extremely large distance r, when the gravitational pull is very tenuous. After all, Newton never dealt with colossal distances of the order of tens of thousands of light years (1 light-year is approximately 5.88 trillion miles) like those represented in Figure 1.8. The extant technology in the 17th century did not enable such observations. In fact, data regression analysis of experimental results of the type presented in Figure 1.9 has led to a modification of Newton's law adapted for huge intra-galactic distances. Thus, the Israeli physicist Mordehai Milgrom has dared to modify Newton's equation in an effort to adapt it to the outpour of astrophysical data. While commendable, his effort is mostly regarded as a formula for data regression, essentially a data-fitting device and not a fundamental advance in the physics underlying galaxy rotation [1].

The physics community overwhelmingly favors the dark matter hypothesis and has chosen to uphold the sanctity of Newton's law [2]. As we shall now see, there are other experiments that support the dark matter hypothesis and favor it over Milgrom's defiance of Newtonian physics [1, 2].

1.6 GRAVITATIONAL LENSING BY DARK MATTER

There is a well-known way of determining the mass of a galaxy, which is actually a feature of Einstein's general relativity called "gravitational lensing." As mentioned in the previous chapter, a massive object such as a galaxy will bend or warp the fabric of space-time, causing a change in curvature, in the form of a dimple (Figure 1.10). Since light travels in space-time, it will be affected by the gravitational pull exerted by the massive object. This interaction between gravity and light elicited huge skepticism when first proposed by Einstein, but ultimately proved to be a decisive

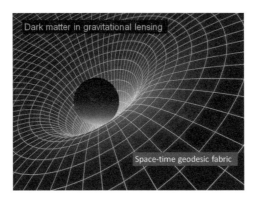

FIGURE 1.10 Relativistic picture of space-time (grid plane) curved or warped by the presence of a massive object.

validation of general relativity, as gravitational lenses were experimentally confirmed. Before dwelling on the mechanism of this phenomenon and leveraging this relativistic feature as a means to detect dark matter we need to prove rigorously that light may be indeed be influenced by gravity. This amazing observation is not in the least obvious since the photons, the "particles" of light are known to be massless.

1.6.1 Gravity Deviates Light: The Power of Thought Experiments

This question is at the core of relativity. To show that light is indeed affected by a gravitational field, we shall resort to Einstein's favorite theoretical tool, the thought-experiment (*Gedankenexperiment* in German). No fancy equipment is required, no expensive gadgets, only the imagination, unbound, tempered only by the principles of physics and their logical consistency. In fact, let us play Einstein for a while. We can imagine young Albert at his modest frugal desk in the patent office at Bern, eyes closed, complete silence, prodding his imagination as he posits his favorite question "What if …?"

Let us imagine a box floating in deep space vacuum (Figure 1.11) with no forces acting on it. A photon of light leaves one wall of the box and travels to the opposite wall at the other side of the box. Let us place the center of coordinates at the point on the wall where the photon started its trajectory. As the photon is absorbed it transfers some modicum of energy that we may denote E (leave your car in the sun and you will have trouble getting back in it). This implies that the photon must carry momentum, that is, what we would identify with "impetus" in common parlance. But the momentum, p, of a particle is usually assessed as $p = mv$, where m

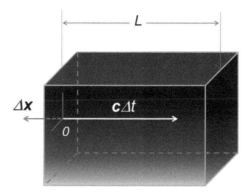

FIGURE 1.11   Representation of the thought experiment that lead Albert Einstein to formulate his famous equation $E = mc^2$.

denotes the mass of the particle and v is its velocity. Since we are told that the photon is massless, how can it possibly have momentum? This question troubled young Einstein for a while. The photon travels at the speed of light, $v = c$, so to compute the momentum of the photon, Einstein used the equation $p = E/c$. This equation does not involve the mass, the troublemaker. As it turned out, that was a clever move to circumvent the problematic mass of the photon, since Einstein knew that the kinetic energy of a moving object traveling at speed v may be computed as $E = mv^2 = pv$, so Einstein simply wrote $p = E/v$, which becomes $p = E/c$ for the photon that travels at the speed of light. This turned out to be a great trick!

Like most of us, Einstein was also familiar with the principle of action-reaction. Everyone who has fired a gun knows this principle: When the gun fires, the bullet goes one way, the gun the other – it's commonly known as recoil or kick. This principle has a fancy name in physics; it is known as the principle of momentum conservation. When we apply it in the context of our thought-experiment, we need to equate the reaction of the box to the movement of the photon inside. Both momenta must be equal in magnitude to compensate each other. Thus, at time Δt from the moment the photon left the wall, the following relation must hold:

$$\frac{M\Delta x}{\Delta t} = \frac{E}{c}, \tag{1.2}$$

where $\dfrac{\Delta x}{\Delta t}$ is the velocity of the box "recoiling" from the photon departure, M is the mass of the box and Δx is the distance travelled by the box at time

Δt, while in that time the photon travelled a distance $c\Delta t$ in the opposite direction (Figure 1.11).

Since the photon travels at the speed of light, the time it takes for it to reach the opposite wall of the box at distance L is $\Delta t = L/c$. Now suppose we replace the photon with a particle with mass. At the time $\Delta t = L/c$, the conservation of momentum would read: $\dfrac{M\Delta x}{\Delta t} = \dfrac{mL}{\Delta t}$, or $M\Delta x = mL$. Using Equation (1.2), we can determine that at time L/c, the box has been displaced by the amount $\Delta x = \dfrac{E\Delta t}{Mc} = \dfrac{EL}{Mc^2}$. Since we already showed that $M\Delta x = mL$, we get $M\dfrac{EL}{Mc^2} = mL$, which is rewritten as $E = mc^2$, Einstein's arch-famous equation!

This means that mass and energy are interchangeable, a result we shall exploit to investigate the nature of dark matter and dark energy. As it is well known, since c^2 is a colossally large quantity, it transpires that a small amount of mass is capable of transmutation into a huge amount of energy. This observation heralded the power of nuclear energy in a weaponizing context. Another consequence of this equation of direct relevance to our previous discussion of gravitational lensing is that since that a particle with mass $m = E/c^2$ may be regarded as a proxy for a photon carrying energy E, light is indeed affected by gravitation, as Einstein correctly predicted.

1.6.2 Gravitational Lensing as a Dark Matter Detector

To understand the principle, let us consider a photon of light (the minimal package of light) that barely skims the boundary of the galaxy. This beam will be deflected at an angle α (Figure 1.12). A general relativity calculation of the deflection of light caused by the gravitational influence of the galaxy gives the rigorous result:

$$\alpha = 4MG/Rc^2. \tag{1.3}$$

In the equation, M is the total mass of the galaxy and R is the radius of the galaxy.

We can also grasp this phenomenon and derive Equation (1.3) from basic physics. Newton assumed that particles were corpuscles, perhaps not endowed with mass but certainly with momentum. We now know that the photon has momentum (E/c) but no mass, yet from the relation $E = mc^2$ we may assume that it is influenced by gravity since mass becomes a proxy for

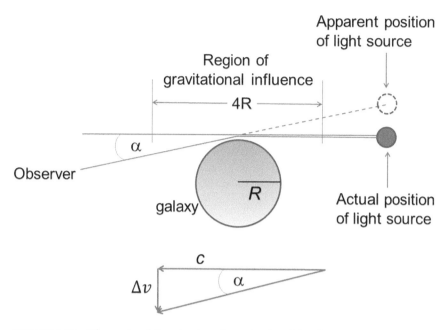

FIGURE 1.12   Elements of the physics of gravitational lensing.

energy, just like Einstein predicted. Adopting this picture, the photon with velocity c (the speed of light) would undergo a deflection of its straight-line trajectory and hence its velocity vector (arrow of magnitude c) will change direction under the gravitational influence of a massive object. This change in direction requires a force, that is, the photon is subject to an acceleration, a, perpendicular to its original direction. The force that deflects the light is simply the familiar $F = ma = \dfrac{GMm}{R^2}$, where M is the mass and R is the radius of the galaxy. This means that the acceleration undergone by the photon of light as it passes by a galaxy is $a = \dfrac{GM}{R^2}$. Since the acceleration is the change in velocity over time, that means that its change in velocity Δv would be given by $\Delta v = a\Delta t$, where Δt is the time interval during which the galaxy exerts a significant gravitational pull. We may assume this force is maximized as light is closest to the galaxy, and since the galaxy radius is R, the force is fully exerted during the travelled distance $4R$, as indicated in Figure 1.12. Note that $4R$ is an enormous distance, of the order of tens of thousands of light-years as illustrated in Figure 1.8. But the photon travels at the speed of light, so that the time interval becomes $\Delta t = \dfrac{4R}{c}$. This means that the perpendicular change in velocity caused by the gravitational pull of the galaxy is $\Delta v = a\Delta t = \dfrac{\left(\dfrac{GM}{R^2}\right)4R}{c} = 4MG/Rc$. So, assuming the deflection angle is small, we may use the approximation

$$\alpha \approx \sin\alpha = \frac{\Delta v}{c} = \frac{4MG}{Rc^2}. \tag{1.4}$$

The formula in Equation (1.4), based on crude assumptions and obtained from basic physics, is *identical* to the one obtained by general relativity (Equation 1.3).

This result demonstrates the astonishing consistency between general relativity and Newtonian physics.

The deflection angle in gravitational lensing of massive galaxies has been measured experimentally and is approximately six times larger than the value obtained from general relativity or classical physics. *Because the deflection angle is proportional to the total mass M of the galaxy, this result implies that the total mass M of the galaxy is about six times larger than the value calculated from the visible and detected matter in the region contained within the sphere of radius R, the radius of the galaxy.* This is a shocking result and implies that the matter that cannot be detected, the dark matter, contributes five times more than the visible matter to the gravitational pull of the galaxy [2]. This result is in striking agreement with the proportion of dark matter to visible matter in the universe given previously: $(26.7){:}5 \approx 5.3$.

1.7 VALIDITY OF THE DARK MATTER HYPOTHESIS

This chapter illustrates the way physics works best. Experimental observations that stand in defiance of a prevailing paradigm are initially invariably regarded as "anomalies" or "systematic errors," and only if they pass the initial peerage scrutiny, they may be taken seriously as worthy contenders of an existing paradigm. This implies that the standard model or laws that are expected to underpin the newly observed phenomena may undergo revision or be completely reformulated in a new guise that often represents a synthesis of two clashing proposals of reality. This narrative describes the saga of dark matter as it unfolded in this chapter. The extant physical models and laws cannot be adapted to encompass the experimental observations in the contexts of galaxy rotation and gravitational lenses, unless the mass responsible for the gravitational pull gets a significant contribution from matter of a hitherto unknown nature. This is precisely dark matter and only an approach bold enough to be unencumbered from the weight of standard models can delineate its nature. As we shall show later in this book, time seems to be ripe for AI to prod over the mass astrophysical data and elucidate the nature of dark matter.

1.8 THE HUBBLE CONSTANT IN THE NEWTONIAN AND RELATIVISTIC UNIVERSE

Evidence that massive amounts of energy of unknown origin keep pouring into the universe without undergoing dilution has been piling up since 1998. That was the year astronomers verified the accelerated expansion of the universe. As space expands, more vacuum forms, and with it more geometrically undiluted energy is infused into space, causing space to expand even further in a sort of autocatalytic reaction where dark matter creation appears to be self-stimulated.

The argument for accelerated expansion runs as follows: Expansion of space stretches light, shifting it to longer wavelengths, hence shifting light toward the red section of the visible spectrum. Light from supernovae appear more "redshifted" the farther away they are from us, because their light has to travel farther through an expanding space. If space expanded at a constant rate, a supernova's redshift would be proportional to its distance to the observer, and thus to its brightness. This is not what astronomers have been observing.

In an accelerating universe filled with undiluted dark energy that scales with vacuum dimensions, space expanded more slowly in the past than it does now. This means a supernova's light will have stretched less during its journey to earth, given how slowly space expanded during much of the time compared with the speed at which it expands now. The light from a supernova located at a given distance away (as determined by its brightness) will appear significantly more redshifted than it would in a universe that lacks dark energy. Indeed, researchers find that the redshift and brightness of supernovae scale in precisely that way. The actual scaling enabled them to compute the amount of dark energy in the universe with significant precision.

At the turn of the 20th century, the cosmological debate had a very different flair. In order to calibrate his ideas as he started developing his general theory of relativity, young Einstein had many conversations with astronomers. They assured him that the universe was essentially static, with no beginning and no end, with galaxies at fixed positions unchanged on a cosmic scale. This picture posed a great problem to Einstein as he sought to reconcile it with the idea that galaxies exerted a gravitational pull on one another that would ultimately lead to a collapse of space-time into a big crunch. Einstein circumvented this problem by introducing a "cosmological constant," a fudge term representing a force that would counter the gravitational pull in order to maintain the stasis of the universe, as it was wrongly related to him by contemporary astronomers [1]. As it turns out, the cosmological constant ultimately relates to the presence of dark energy but not in any way remotely resembling what Einstein would have anticipated.

Fortunately, before Einstein went too far with the cosmological constant idea, an American astronomer by the name of Edwin Hubble (1889–1953) came along and showed that, in fact, the universe was expanding!

Expansion means that the distance between any two points in the universe increases with time but it does so in relation to a single time-dependent scale factor that we shall denote $a = a(t)$. It is crucial to note that the scale factor is the same in all space and only varies with time. In rigorous terms, the expansion idea may be formalized as follows: Let $d_{ij}(0)$ be the distance between any two points labelled "i" and "j" in the universe at a particular time conventionally set to be $t = 0$, then the distance $d_{ij}(t)$ at a future time t becomes: $d_{ij}(t) = d_{ij}(0).a(t)$ and this equation holds for any pair i, j of points in space. The important thing is that there is only one scaling factor and it is not dependent on the pair of points chosen. Hubble noticed that this scaling factor increases with time.

As the concept settles down in our minds, let us drop subindices and adopt a more agile notation, denoting $y = y(t) = d_{ij}(t)$ and $\Delta y = d_{ij}(0)$. Then, $y(t) = \Delta y.a(t)$ and dy/dt, the rate at which distance y changes in time, becomes $dy/dt = \Delta y.(da/dt)$, where da/dt is the rate of change of the scaling factor of the universe. Combining the previous relations we may write:

$$\frac{dy}{dt} = \Delta y \frac{da}{dt} = y \left[\frac{\left(\dfrac{da}{dt}\right)}{a} \right]. \tag{1.5}$$

The factor $\left[\dfrac{\left(\dfrac{da}{dt}\right)}{a} \right] = H$ in the left-hand side is the Hubble constant of the universe, that is, the rate at which the scale factor changes as a result of the universe expansion divided by the scale factor [1]. It is important to point out that H is constant in space but not in time. Equation (1.5) has far-reaching consequences: If we know the velocity $\dfrac{dy}{dt}$ of a receding galaxy, we can calculate y, that is, how far away it is. Furthermore, as we shall show later in the discussion, *the more distant the galaxy is from us the faster it is traveling away from us.*

This is the Hubble picture of the expanding universe, and in this picture, space itself is expanding. It has to, implying that vacuum is getting larger and larger, a troubling runaway scenario by no means well understood.

Let us elaborate this context further by applying the well-worn Newton's law of gravitation. The gravitational pull exerted on a galaxy revolving at a distance y from an arbitrarily chosen point in the universe would be $F = GMm/y^2$, where M is the total mass of matter contained within the ball of radius y centered at the chosen point, and m is the mass of the revolving galaxy. The potential energy (U) of the galaxy is the physical magnitude whose rate of change in time is the gravitational force, therefore $U = -\,GMm/y$. Now, the total energy (E) of the galaxy must include the potential energy associated with the gravitational force exerted on it plus another contribution (K) that represents the kinetic energy, that is, the energy associated with the movement of the galaxy. Let us assume the latter form of energy is only associated with the expansion of the universe. Then, we may write $E = K + U = \dfrac{1}{2}m\left(\dfrac{dy}{dt}\right)^2 - GMm/y$. But E is unchanged in time unless another force starts working on the galaxy, which we assume not to be the case. So, we may say that the quantity $\left[m\left(\dfrac{dy}{dt}\right)^2 - 2GMm/y\right]$ is a constant or $\left[\left(\dfrac{dy}{dt}\right)^2 - 2GM/y\right]$ is constant, since the mass m remains unchanged.

Now, plugging in the previously obtained Equation (1.5) into the last expression, we obtain:

$$y^2\left(\frac{da/dt}{a}\right)^2 - \frac{2GM}{y} = \text{constant.} \tag{1.6}$$

The mass enclosed by the sphere of radius y is $\dfrac{4}{3}\pi y^3 \rho(t)$, where $\dfrac{4}{3}\pi y^3$ is the volume of the ball of radius y and $\rho(t)$ is the mass density that obviously keeps decreasing in time in an expanding universe. We may then write:

$$y^2\left(\frac{da/dt}{a}\right)^2 - \frac{2G}{y}\frac{4}{3}\pi y^3\rho(t) = \text{constant.} \tag{1.7}$$

A little algebra (division by y^2) and substitution with known relations derived above yields:

$$\frac{\left(\frac{da}{dt}\right)^2}{a^2} = H^2 = \frac{8\pi G}{3}\rho(t) - k/a^2. \tag{1.8}$$

In this equation, k on the right-hand side is a constant. Equation (1.8) is known as the Friedman-Robertson-Walker (FRW) formula derived from the theory of general relativity [1]. But we have shown that a simplified version of the formula could be obtained from Newtonian physics, making very simple assumptions.

The reader is once more reminded that H in Equation (1.8) stands for the Hubble constant, which is a constant in space and not in time. Notice that generally the following relation holds: $\frac{8\pi G}{3}\rho(t) - k/a^2 \geq 0$. This relation becomes the hallmark for an "open universe." On the other hand, if at some point in time we get $\frac{8\pi G}{3}\rho(t) < k/a^2$, the universe would stop expanding and start contracting and we would call it a "closed universe." In this case the universe would eventually collapse onto itself into a singularity under the gravitational pull. This scenario has been termed the "big crunch." If on the other hand $k = 0$, the universe will expand at an ever-decreasing rate because the matter density is monotonically decreasing as the universe keeps expanding. The rate of decelerated expansion will approach the asymptotic limit value zero at infinite time. This type of universe is called a "flat universe." The time dependence of the size of the three types of universe – open, closed and flat – is schematically shown in qualitative fashion in Figure 1.13.

1.9 THE EXPANDING UNIVERSE REVISITED

The best up-to-date observations of our skies support the picture of a flat universe ($k = 0$), that is, it will continue to expand only until it reaches

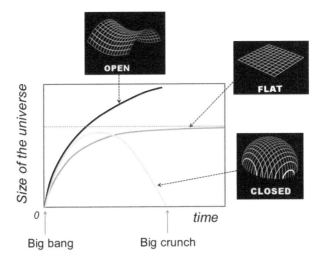

FIGURE 1.13 Qualitative scheme of the time dependence of the size of the universe for a closed universe (light grey), open universe (black), and flat universe (grey).

a particular size. To compute the density ρ, let us assume a cubic box of dimension a in a flat universe containing mass M. Then, the FRW formula becomes

$$\frac{\left(\dfrac{da}{dt}\right)^2}{a^2} = H^2 = \frac{8\pi G}{3}\frac{M}{a^3}.$$ (1.9)

This equation can be simplified to $\dfrac{da}{dt} = \sqrt{\dfrac{8\pi GM}{3a}}$, which can be rewritten as

$$\left(da/dt\right) = Wa^{-1/2}.$$ (1.10)

In Equation (1.10), the constant W is defined by parameters of the universe: $W = \sqrt{\dfrac{8\pi GM}{3}}$. Now, to find out how the universe expands in time, we need to compute the time dependence of $a = a(t)$. This may be simply obtained by noting that Equation (1.10) may be rewritten as $a^{1/2}da = Wdt$ giving $a^{3/2} = w' t$, or equivalently, we may state that our universe is expanding as

$$a = wt^{2/3},$$ (1.11)

where *w' and w* are constants. This is the *predicted* expansion behavior for our current matter-dominated flat universe. As we shall see, the experimental results reveal a huge discrepancy with this rigorously obtained prediction, and dark energy is at the core of the problem!

At the beginning of its existence, right after the big bang, the universe would not have been matter-dominated but rather radiation-dominated, with spontaneous creation of matter and its compensatory antimatter that would mutually annihilate with a huge production of photons. A residual amount of matter would have been created as a proxy for the energy released, as described by Einstein's formula $E = mc^2$, derived in the previous chapter. In this early universe, let us consider a cube of photons of dimension *a*. How would this scale factor for the expansion in the early radiation-dominated universe behave in time?

To describe this early universe we need to resort to some basic principles of quantum mechanics. We have previously discussed that the photon is a massless particle and yet it carries momentum. This strange duality was something that Einstein found particularly irksome because he could not reconcile that fact with the fact that the photon transfers energy as it hits a surface. He eventually resolved the paradox admirably, as discussed previously Einstein's results in fact imply that the notion of particle needs to be refined. We can no longer think of an elementary particle as a "corpuscle" but rather as a "wave-like excitation" or a "warp in a field." The photon travels at the speed of light but its "kinetic energy" depends on how many crests are packed in the wave per unit time; this is what we know as frequency. Max Planck showed that in fact the energy of the photon may be written as $E = hf$, where *f* is the frequency and *h* is Planck's constant. In fact, *h* is one of the parameters for our universe, together with G, Newton's gravitational constant, and a few other constants. Together they define how our universe behaves.

Planck's expression for the energy of a photon may be rewritten as $E = hc/\lambda$, where c is as usual the speed of light and λ is the wavelength, that is, the distance between two consecutive crest peaks of the "wave-particle." The equivalence between both expressions for the energy of the photon follows from the simple fact that, by definition, $f = c/\lambda$. As *a* increases with the universe expansion, so does everything else, including the wavelength of the photon that becomes commensurate with *a*. The photons in the early universe have their wavelengths stretched as the universe expanded, implying a dramatic cooldown following the big bang, since the energy goes down as the wavelength increases (the frequency decreases). The wavelength of the "cold" ancient photons is today in the microwave region, a very low-energy region of the radiation spectrum. This is the "glow" we see in the skies known as Cosmic Microwave Background Radiation (CMBR), discussed previously. The CMB is as it were the relic of the early radiation-dominated universe that followed right after the big bang.

Since the wavelength of the ancient photons can be made proportional to the expansion factor in the radiation-dominated early universe, we may write $E = J/a$, where J is a constant. This gives a radiation energy density $\rho_r = \left(\dfrac{J}{a}\right)\left(\dfrac{1}{a^3}\right) = J/a^4$. The a-dependence in this expression can be contrasted with the previously obtained density for a matter-dominated "later" universe: $\rho = M/a^3$.

We can substitute the expression for the density in the radiation-dominated universe in the FRW formula given by Equation (1.8), keeping in mind that our universe is flat and hence $k = 0$. This substitution gives

$$\frac{\left(\dfrac{da}{dt}\right)^2}{a^2} = \frac{8\pi GJ}{3a^4}. \tag{1.12}$$

Equation (1.12) can be rewritten as $ada = Cdt$, where $C = \sqrt{8\pi GJ/3}$, yielding the following time dependence for the expansion in the radiation-dominated early universe: $a \propto t^{1/2}$. This behavior is different from the expansion of the matter-dominated universe, which, as we may recall, has a time dependence of the form $a \propto t^{2/3}$.

The two predicted modes of expansion of the universe can be best visualized in qualitative graphs. Thus, Figure 1.14 shows density versus expansion scale factor a for the radiation-dominated and the matter-dominated universes. There is a crossover point at approximately 10 thousand years from the big bang, since the density of the radiation-dominated universe falls more steeply than the density of the matter-dominated universe as a increases. Alternatively, we may plot size of the universe against time (Figure 1.15). The radiation-dominated universe size increases more slowly (as $t^{1/2}$) than the matter-dominated curve with $a \propto t^{2/3}$. So the latter curve crosses the former at the point determined previously to be in the order of 10 thousand years. As previously discussed, the expansion curve for the matter-dominated universe, that is, the universe we live in today, will reach an asymptotic limit. This prediction, as we shall see subsequently, is at odds with experimental observations that reveal an accelerated expansion of the

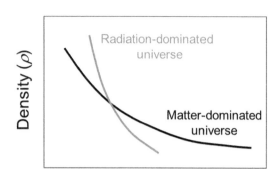

FIGURE 1.14 Qualitative behavior of density in an expanding radiation-dominated and matter-dominated universe.

universe. This discrepancy between prediction and experiment will be shown to be at the core of the dark energy controversy.

In summary, general relativity (FRW equation) predicts that the universe has expanded under radiation dominance until about 10K years and then it expanded faster under matter dominance and that this expansion will slow down to reach an asymptotic limit at infinite times, as described in Figure 1.15.

As we keep anticipating, this is not what seems to be happening according to experimental evidence. The universe expansion is not showing signs of slowing down. Quite the contrary, the expansion is accelerating, prompting us to invoke the presence of a form of energy of unknown origin that is fueling the runaway. This energy of unknown nature has been

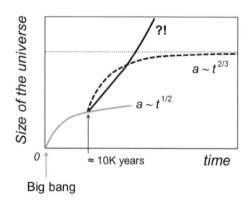

FIGURE 1.15 Qualitative behavior of the time dependence of the size of the universe in a radiation-dominated regime (grey line), matter-dominated regime (dashed line), and experimentally validated runaway regime (dark solid line).

ominously termed "dark energy" and may well become the nemesis of our lofty theories on how the universe works [2].

1.10 AUTOCATALYTIC UNIVERSE IN RUNAWAY MODE

The accelerated expansion of the universe constitutes the biggest challenge to the prevailing paradigm in physics, as it demands either a complete revision of the physical laws to account for this "anomaly", or alternatively – but no less painfully – that we reckon the existence of dark energy, a form of energy of hitherto unknown origin that is constantly being injected into the universe and is causing its runaway behavior in complete defiance of gravitational laws.

The runway picture of the universe was put forth by Perlmutter, Schmidt, and Riess, as discussed previously [1]. They were able to establish the exponential expansion by precisely measuring the so-called redshift effect in the radiation collected from very bright objects called *supernovae*. As the light-emitting source travels away from the observer, the detected frequency f of the emitted light gets lower because the time of arrival of successive wave crests gets longer (Figure 1.16). The net effect can be represented as a stretching of the wave. The observed frequency f is related to the emitted frequency f_e by the relation: $f = f_e \left[\dfrac{c}{c + v_s} \right]$, where v_s is the

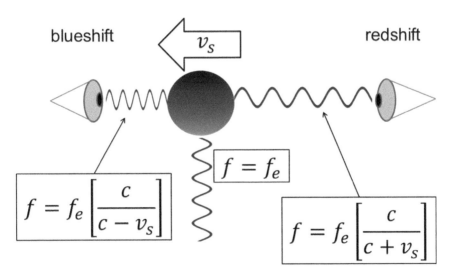

FIGURE 1.16 The redshift and blueshift effect on observed radiation from a light-emitting source moving with velocity v_s relative to the observer.

velocity of the light source. The apparent stretching of the wavelength is known as redshift since the red color of the visible spectrum corresponds to the longest wavelength and lowest frequency. Thus, the redshift is implied by the relation $f < f_e$.

Conversely, if the light-emitting source approaches the observer at speed v_s, the time of arrival of successive waves is increased, so the waves are bunched together, compressed as it were, with a net "blueshift" effect resulting in increased frequency ($f > f_e$) in accord with the formula

$$f = f_e \left[\frac{c}{c - v_s} \right]. \tag{1.13}$$

The accelerated expansion of the universe was established by determining the commensurability of the redshift effect with the brightness of the light-emitting source, a proxy for the distance to the observer on earth. It was determined that the brighter the source, the smaller the redshift, and conversely, the dimmer the source, the more pronounced the redshift (Figure 1.17). The implications of this observation remain as troublesome as they are transparent: The more distant the source, the higher the velocity at which it is traveling away from the observer; in other words: $v_s = v_s(d) \propto d$, where d is here the distance to the light-emitting source. But this implies that $da/dt \propto a$. The solution of this equation yields an exponential or runaway time-dependence: $a \sim e^{qt}$, with $q = $ constant.

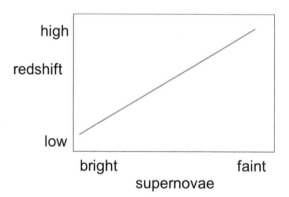

FIGURE 1.17 Qualitative behavior of redshift for nearby and distant light sources. Brightness is accepted to be a proxy for distance. Thus, far away objects move away from the detector at faster speed than objects nearby. This behavior constitutes the hallmark of the runaway universe as validated in the Perlmutter-Schmidt-Riess experiments.

This means that the scale factor a is growing exponentially in time, completely at odds with the predicted scenarios of general relativity described previously. We are in the presence of the hallmark for a runaway universe!

1.11 THE UNIVERSE RUNAWAY FUELED BY DARK ENERGY

In manner of digression, we must consider here alternative ways of computing the energy E of the photon. We have dealt with energies but so far we have not incorporated heat or temperature into our physical picture of the universe. Temperature is the physical quantity conjugated to energy in thermodynamics, the science concerned with the interconversion of heat, energy, and work, whose modern foundations were laid out by the Austrian physicist Ludwig Boltzmann (1844–1906).

It is estimated that the photons in the CMB correspond to radiative emissions of an extremely cold black body, currently at a temperature of approximately 3 degrees Kelvin ($-454.27°F$ or $-270.15°C$). On the other hand, the energy of a photon may be related to the temperature of emission through Boltzmann's formula $E = kT$, where k is Bolzmann's constant. Now, given since $E = hf = hc/\lambda \approx hc/a$, we may conclude that

$$T = \left(\frac{hc}{k}\right)\left(\frac{1}{a}\right). \tag{1.14}$$

This implies that the universe is cooling down as it is expanding. We also have an estimation $T \approx 3000°K$ for the temperature of the universe at a particular epoch after the big bang, namely the ionization epoch, when atoms were stripped of their electron shells, with the atomic nuclei embedded in a plasma of electrons. Thus we may compute the ratio $a_{today}/a_{ion} = 3000/3 = 1000$, where a_{today}, a_{ion} are respectively the expansion scales today and at the ionization epoch. This means that our universe is a thousand times bigger now than at that hot era right after the big bang.

Now, how far back did the ionization take place? We can easily answer this question since we know the time dependence of a in a matter-dominated universe and the current age of the universe ($t = 13.8$ billion years, counted since the big bang):

$$\frac{a_{today}}{a_{ion}} \approx 1000 = \left(\frac{t}{t_{ion}}\right)^{\frac{2}{3}} = \left(13.8\times10^9 \, y/t_{ion}\right)^{2/3}. \tag{1.15}$$

This implies that the ionization era occurred at the time $t_{ion} \approx 460{,}000$ years after the big bang.

Here we reproduce the FRW formula for our flat universe but in a way that accommodates the radiation-dominated regime ($\rho \sim 1/a^4$) as well as the matter-dominated regime ($\rho \sim 1/a^3$):

$$H^2 = \frac{\left(\dfrac{da}{dt}\right)^2}{a^2} = \frac{8\pi G}{3}\rho(t) = \left(\frac{8\pi G}{3}\right)\left(\frac{C}{a^{3(b+1)}}\right). \tag{1.16}$$

In this equation C is a constant and $b = 1/3$ for the radiation-dominated regime and $b = 0$ for the matter-dominated regime.

As we have anticipated, *these regimes postulated by the theory of general relativity do not agree with the experimentally verified exponential expansion of the universe determined by Perlmutter, Schmidt, and Riess* (see previous discussion). These scientists have shown that a behaves exponentially in time: $a \sim e^{\theta t}$ with $\theta = $ constant. Now, from Equation (1.16) it follows that the case where $b = -1$ corresponds to a constant density $\rho = C = \rho_0$. *Paradoxically, the $b = -1$ regime not anticipated by general relativity is precisely the type of universe whose expansion was measured by Perlmutter, Schmidt, and Riess.* By plugging in $b = -1$ in Equation (1.16), we get

$$\frac{\left(\dfrac{da}{dt}\right)^2}{a^2} = \frac{8\pi G}{3}\rho_0 \approx \Lambda, \Lambda = \text{"cosmological constant of the universe"}. \tag{1.17}$$

This formula yields the exponential expansion of the universe, with the fudge factor in Einstein's theory, namely the ill-fated "cosmological constant." We may rewrite it as

$$\frac{da}{dt} = a\sqrt{\Lambda}. \tag{1.18}$$

Equation (1.18) is clearly indicative of a runaway process: The larger the expansion, the larger the rate of expansion. As expected, the equation's solution yields a factor a that is exponentially increasing in time:

$$a \propto e^{\sqrt{\Lambda}t}, \theta = \sqrt{\Lambda} = \sqrt{\frac{8\pi G\rho_0}{3}}. \tag{1.19}$$

In other words, by taking $b = -1$ or fixing the density as constant, we can reproduce the experimental result of exponential expansion and therefore accelerated expansion of the universe that earned Perlmutter, Schmidt, and Riess the Nobel prize in physics. Furthermore, we can calculate the cosmological constant, bestowing physical meaning to this troublesome factor that Einstein regarded as a fudge term ("my biggest blunder," he called it) in his theory of general relativity.

On the other hand, we know that the speed $v = dy/dt = \Delta y.da/dt$ at which a galaxy runs away from us increases as the distance $\Delta y.a$ to the galaxy increases. But this is precisely what Equation (1.18) is telling us: $v = \dfrac{dy}{dt} = \Delta y \dfrac{da}{dt} = \Delta y a \sqrt{\Lambda} = y\sqrt{\Lambda}$ or $v \propto y$. Furthermore, the equation is also telling us that the proportionality factor is the root square of the cosmological constant!

One important consequence of this result is that beyond a critical distance $y^* = c/\sqrt{\Lambda}$, the speed of a galaxy at distance $y > y^*$ will be larger than the speed of light: $v > c$! This means that such a galaxy will travel out of sight: Its light will never reach us. The galaxy will simply disappear beyond our horizon. The accelerated expansion of the universe will make galaxies disappear beyond our horizon, hence heralding a much duller view of the skies.

We may have lifted the veil of nature, but only to unravel yet another mystery. If ordinary matter ($b = 0$) or ordinary radiation ($b = 1$) are not dominant in the universe, what is causing the accelerated expansion of our universe? A mysterious energy of unknown origin is being constantly created and injected into our universe from a source that does not get diluted as the universe expands. This mysterious energy fueling our universe runaway is what we call dark energy.

At least we know one key thing about dark energy: It does not get diluted as the universe expands, since its density is constant ($\rho = \rho_0$). Now, the thing that is constantly being created without getting diluted in the universe as it expands is … vacuum! Indeed, vacuum does not get geometrically diluted as the universe increases its volume, so dark energy is generated by the vacuum! And we know this is not a negligible contribution: The dark energy density ($\rho = \rho_0$) makes up for about 68.3% of the total density of energy/matter in the universe.

1.12 QUANTUM VACUUM FLUCTUATIONS FUEL A RUNAWAY UNIVERSE

Relativity is taking us to an unchartered territory far more rarified even than that of dark matter. The universe expansion is autocatalytic, it has a retro-feeding mechanism for acceleration whereby the more it expands the more it harvests dark energy, which in turn fuels more expansion, creating a runaway out of vacuum generation.

The universe runaway is the inevitable truth that emerges from the current physics paradigm. The self-stimulated vacuum energy creation becomes a major hurdle in our trend of thought and prompts us to prod the complementary theory to general relativity: Quantum mechanics.

To unravel how the creation of dark energy may come about, we refer to quantum field theory (QFT), where an electron and its antiparticle, the positron, occasionally emerge spontaneously in the vacuum with emission of a photon. The particles exist for a brief time and then annihilate each other without a trace unless another event elsewhere in the vacuum leads to an interaction with a component of the triad. Energy conservation is violated, but only for the extremely ephemeral lifetime of the particle Δt permitted by the so-called *uncertainty principle* that posits that $\Delta E \times \Delta t \sim \frac{\hbar}{2}, \hbar = h/2\pi$, where ΔE and Δt represent respectively *a priori* uncertainty in energy and time. This principle is one of the milestones of quantum mechanics and asserts that uncertainties in conjugated quantities like energy and time, or position and momentum, balance each other, yielding a constant product. Thus, a short uncertainty in time may yield a huge fluctuation in energy, utterly commonplace in QFT. There is no limit placed by the laws of physics on the scale of this energy fluctuation. Nothing prevents it from occurring at a grand scale. The duration is of course subject to the restriction of the uncertainty principle, which merely implies that the universe has a zero total energy, which made the fluctuation possible in the first place. Quantum theory implies that the vacuum should be unstable against large fluctuations in the presence of a long-range, negative potential energy term. Gravitation is precisely such a term. This fact encourages us to believe that vacuum creation stimulates further vacuum creation as dark energy adopts the form of vacuum energy fluctuations.

Because vacuum does not undergo any dilution as the universe expands, quantum vacuum energy becomes the prevailing proxy for dark energy, although this answer is far from satisfactory and prompts further investigation, as shown in Chapter 5.

The universe runaway scenario experimentally validated by Perlmutter, Schmidt, and Riess in their Nobel-deserving work is at least compatible with the principle of autocatalytic vacuum creation (AVC) hereby put

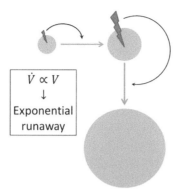

FIGURE 1.18   Kinetic scheme for the principle of autocatalytic vacuum creation (AVC) as the mechanistic underpinning of a runaway universe.

forth (Figure 1.18). The AVC principle can be defined as follows: A quantum vacuum fluctuation fuels the creation of a bigger vacuum, which in turn has an enhanced chance to spontaneously generate a larger quantum fluctuation, which in turn fuels the creation of a bigger vacuum, so the rate of vacuum volume creation $\dot{V} = \dot{V}(t) = \dfrac{dV}{dt}$ at a given time t is proportional to the vacuum volume $V = V(t)$ created at that time. This assertion may be written as $\dot{V} \propto V$. This equation implies a runaway in vacuum creation, as the solution to the equation yields $V \sim e^{Qt}$, where Q is some constant whose relation to the cosmological constant Λ would need to be established.

For now, we are contented with the assertion that the relativistic Equation (1.17) together with the AVC principle provide the physical underpinnings for the runaway universe as fueled by dark energy. Furthermore, this dark energy is expected to be generated in autocatalytic cycles of vacuum creation that are causative of the universe runaway behavior.

1.13 DARK PHYSICS ON EXTRA DIMENSIONS

A survey of the evidence described in the previous sections prompts us to question whether a shift in paradigm is in order to encompass deep space phenomenology. It appears that the influence of dark matter and dark energy cannot be done with by merely patching up the laws of physics or extending them to include cosmic ranges of interaction hitherto uncalled for. So far, the discussion has included local physical theories, that is, theories whose predictive power resides in solving differential equations or become appropriately stylized within the setting of differential geometry. Topological attributes of the universe have not entered into the equations in any obvious manner and the topology of the universe has not been so far a matter of concern (neither it was in Einstein's time). However, as argued in this book, topology plays a major role in elucidating the nature of dark matter and dark energy. The fact that there is a form of matter in vast proportion that is not associated with visible light quanta calls for an extra dimension, as shall be argued in the forthcoming chapters. The main hurdle to incorporate an extra dimension is topological, not geometrical.

There have been early attempts by Theodor Kaluza and Oskar Klein (KK theory, revisited in [6]) to incorporate an extra compact curled-up dimension into what was otherwise regarded at the time as a "Euclidean universe." These attempts held promise at least initially and were taken seriously by none other than Albert Einstein himself, who spent the latter part of his life seeking to unify gravity with the other forces of nature. However, a lack of topological understanding of space-time at the time made it impossible to properly incorporate extra dimensions and reconcile these extensions with the data available, especially with data pertaining to the prediction of the rest mass of elementary particles.

In this book the "KK theme" may be said to be resurrected albeit in a vastly different guise and for a completely different purpose. First, we shall show in the forthcoming chapters that the universe is compact, regardless of its perceived or inferred extension, and multiply connected, and that these topological attributes determine the way the extra spatial dimension gets incorporated. The extra dimension becomes a necessity to explain dark matter and dark energy, as can be deduced from the simple argument laid out subsequently.

Suppose we consider a spatial cross-section of the universe along a single compact dimension with radius r. We can store energy in this dimension by considering a stationary "de Broglie" wave with wavelength $n\lambda = 2\pi r$, and we know that the kinetic energy of the wave, $E = \dfrac{hc}{\lambda} = \dfrac{\hbar nc}{r}\left(\hbar = \dfrac{h}{2\pi}\right)$ can be translated into a mass: $m = \dfrac{\hbar n}{cr}$.

Now, assume there is an extra "dormant" dimension in space with radius r' that stores dark matter (Figure 1.19). This is a fundamental tenet in the theoretical development put forth in this book and is presented here in a simplified fashion for the sake of the argument. The key difference

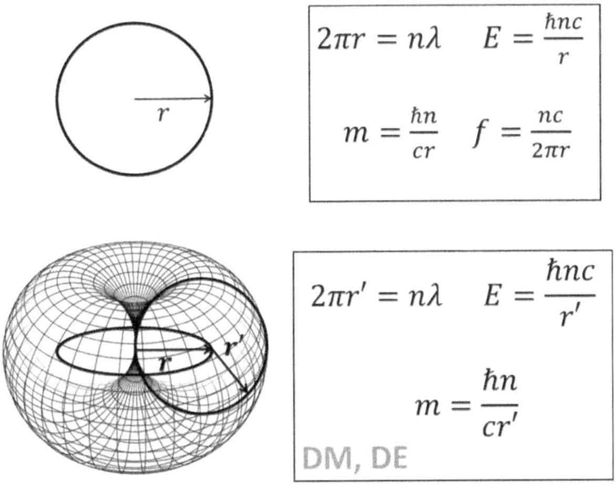

$$2\pi r = n\lambda \qquad E = \frac{\hbar nc}{r}$$

$$m = \frac{\hbar n}{cr} \qquad f = \frac{nc}{2\pi r}$$

$$2\pi r' = n\lambda \qquad E = \frac{\hbar nc}{r'}$$

$$m = \frac{\hbar n}{cr'}$$

DM, DE

FIGURE 1.19 Dark matter/energy stored along a dormant dimension that cannot be associated with detectable radiation.

with matter waves along detectable dimensions is that the mass $m = \frac{\hbar n}{cr'}$ cannot be associated with radiation at the Einstein frequency $f = \frac{nc}{2\pi r'}$.

In other words, *the validity of the relation $E = mc^2$ is contingent on having radiation and its associated matter stored on the same set of observable dimensions*. A staggering but inevitable consequence of this trend of thought is that Einstein's relation is only fulfilled by 5% of matter in the universe.

1.14 THE DIRAC EQUATION: A FIRST GLIMPSE AT VACUUM ENERGY

It is imperative that we discuss vacuum energy in our quest to investigate the nature of dark energy. The natural starting point for such a discussion is the Dirac equation, a breakthrough that inspired an enormous intellectual effort, quantum field theory, whereby vacuum energy was properly interpreted and cast in a rigorous footing.

The first successful synthesis between special relativity and quantum mechanics was accomplished by Paul Dirac (P. A. M. Dirac, 1902–1984) with his relativistic wave equation for the electron conceived in 1928 [7]. Dirac's strategy was to adopt Einstein's relativistic equation

$$E^2 = \left(pc\right)^2 + \left(mc^2\right)^2, \tag{1.20}$$

with m = rest mass, p = momentum, as a starting point and replace the physical quantities for their corresponding operators in consonance with the Schrödinger equation. At a variance with the latter, Dirac's wave function becomes a four-component vector, aptly named a *spinor*, where spin or intrinsic angular momentum emerges as a conserved quantity associated with a quantum number. To yield meaningful solutions, Dirac "linearized" Equation (1.20) by taking the root square: $E = \pm\sqrt{(pc)^2 + (mc^2)^2}$, essentially "factorizing" the relativistic equation. The existence of negative energies proved extremely problematic and initially exposed Dirac to severe criticism from his peers. Dirac considered the momentum as a four-component vector $\vec{P} = \left(\dfrac{E}{c}, P_x, P_y, P_z\right) = (P_0, P_1, P_2, P_3)$ where the dot product is determined by the metric $g_{\mu\nu} = \eta_{\mu\nu}$ given in Figure 1.20. Einstein's convention of summation over repeated indices at different levels is followed throughout the book.

$$\| \vec{P} \|^2 = P_\mu.P^\mu = \eta_{\mu\nu}P^\mu P^\nu = \left(\frac{E}{c}\right)^2 - \left(P_x^2 + P_y^2 + P_z^2\right); \qquad (1.21)$$

$$P_\mu = g_{\mu\nu}P^\nu. \qquad (1.22)$$

To simplify notation, we adopt a unit of velocity equal to the speed of light, so that $c = 1$. Thus, the relativistic Equation (1.20) reads

$$P_\mu.P^\mu - m^2 = 0. \qquad (1.23)$$

Dirac was able to factorize Eqation (1.23) as a wave equation by substituting momentum components for operators ($P_0 = E \rightarrow i\partial_t$, $P_1 \rightarrow -i\partial_x$,

$$\eta = \begin{pmatrix} -1 & 0 & 0 & 0 \\ 0 & 1 & 0 & 0 \\ 0 & 0 & 1 & 0 \\ 0 & 0 & 0 & 1 \end{pmatrix}$$

FIGURE 1.20 A Minkowski metric for 4-dimensional space-time with inner product metric signature −+++.

$P_2 \rightarrow -i\partial_y, P_3 \rightarrow -i\partial_z; \hbar = 1$) as prescribed by the Schrödinger formulation, yielding the relativistic wave equation for the electron in the form:

$$\left(i\gamma^\mu\partial_\mu - m\right)\psi = 0 \tag{1.24}$$

with the four (contravariant) gamma matrices indicated in Figure 1.21 and the wave function now represented as a four-component vector called *spinor*. The gamma matrices satisfy anticommutative constraints imposed by comparing Equations (1.23) and (1.24):

$$\left\{\gamma^\mu, \gamma^\nu\right\} = \gamma^\mu\gamma^\nu + \gamma^\nu\gamma^\mu = 0 \text{ when } \mu \neq \nu; \gamma^\mu\gamma^\mu = 2g^{\mu\mu}. \tag{1.25}$$

The wave function for a particle at rest is

$$\psi = \left(\psi_1, \psi_2, \psi_3, \psi_4\right)^T = \left(a_1, a_2, a_3, a_4\right)^T e^{-iEt}; \sum_{j=1,\ldots,4} a_j^2 = 1. \tag{1.26}$$

Thus, its first two spinor components require that $E = m$, the expected result since $c^2 = 1$ in our convention. However, the last two spinor components yield solutions with the "unrealistic" $E = -m$. In states of a moving electron defined by nonzero third and fourth spinor components, we obtain $E = -\sqrt{(p)^2 + m^2}$, implying that the electron would tend to constantly increase momentum and emit radiation to reach lower and lower energy, all the way to $E = -\infty$, clearly an absurdity that elicited harsh criticism toward the Dirac approach.

Dirac addressed this problem by introducing the following assumptions:

- Dirac's equation is actually of broader applicability, as it holds for any fermion with spin ½. This is because of the way in which the gamma matrices are constructed based on Pauli's spin matrices (Figure 1.21): $\gamma^j = i\sigma^2 \otimes \sigma^j; \gamma^0 = \sigma^3 \otimes \mathbb{I}_2; \otimes = $ Kronecker product.

- The relativistic wave equation for the electron is actually an equation for **all** electrons in the universe.

- All or most states with negative energy are occupied by electrons/fermions, and Pauli's exclusion principle comes into play, forcing the detectable electron/fermion, describable with Dirac's equation,

$$\gamma^0 = \begin{pmatrix} 1 & 0 & 0 & 0 \\ 0 & 1 & 0 & 0 \\ 0 & 0 & -1 & 0 \\ 0 & 0 & 0 & -1 \end{pmatrix} \qquad \gamma^1 = \begin{pmatrix} 0 & 0 & \boxed{0} & \boxed{1} \\ 0 & 0 & \boxed{1} & \boxed{0} \\ 0 & -1 & 0 & 0 \\ -1 & 0 & 0 & 0 \end{pmatrix} \sigma_1$$

$$\gamma^2 = \begin{pmatrix} 0 & 0 & \boxed{0} & \boxed{-i} \\ 0 & 0 & \boxed{i} & \boxed{0} \\ 0 & i & 0 & 0 \\ -i & 0 & 0 & 0 \end{pmatrix} \sigma_2 \qquad \gamma^3 = \begin{pmatrix} 0 & 0 & \boxed{1} & \boxed{0} \\ 0 & 0 & \boxed{0} & \boxed{-1} \\ -1 & 0 & 0 & 0 \\ 0 & 1 & 0 & 0 \end{pmatrix} \sigma_3$$

FIGURE 1.21   Contravariant Dirac matrices.

into a state with positive energy. The negative-energy electrons are completely unobservable, and its collection became known as the "*Dirac sea.*"

- An electron (or 1/2–spin fermion in general) in the highest negative energy level ($E = -m$) can become "detectable" by absorbing a photon of energy $E = hf = 2mc^2$. Such a photon can cause an electron to transition from the highest negative energy state ($E = -m$) to the lowest positive energy state ($E = m$). Half of the energy absorbed would correspond to the emergence of a detectable electron and the other half would correspond to a "hole" in the Dirac sea, and since a negative charge has been taken away into the detectable realm, the hole is equivalent to a positive "antielectron," which in modern terms would be named "positron."

These assumptions lead to predict the existence of antimatter, a prediction eventually confirmed by experiment. Positrons were detected by Nobel laureate Carl David Anderson (1905–1991) within the spontaneous formation of electron-positron pairs in a cloud chamber.

1.15 NAVIGATING THE DIRAC SEA TOWARD DARK ENERGY

The Dirac sea was ultimately considered a dated construction superseded by quantum field theory (QFT). In QFT, particles are considered not describable individually or even in a closed system but rather excitations of universal fields. Furthermore, the emergence of the electron-positron

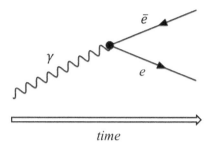

time

FIGURE 1.22 Electron-antielectron pair creation in quantum field theory as schematically represented in by its Feynman diagram.

pair is regarded as a spontaneous elementary process in absolute vacuum (Figure 1.22), whereby the detectable fermion spontaneously emerges from vacuum through communication with a photon with a lifetime determined from Heisenberg's uncertainty principle: $\tau \sim \dfrac{\hbar}{4mc^2}$.

With the discovery of the dark universe, the Dirac sea may need to be summoned once again, as it will be described in this book. A fermionic sea totally unaccounted for may be just what is needed to delineate the nature of the dark universe, except that the weak force governing the decay of such particles would not be operative in the way it is described in the Dirac equation. It is well established that dark energy and dark matter cannot be turned visible through photon absorption, so a vast proportion of the negative-energy fermionic sea must be rendered impervious to communication of the electroweak force, but how? The identification of the purveyor to the dark universe is described in Chapter 5 and requires a careful discussion of "chirality," an essential concept in particle physics.

It is obviously tempting to identify the Dirac sea of negative-energy fermions with the dark universe, but this assertion is incorrect and would require serious revision and fine-tuning vis-à-vis the fact that the electroweak interaction that enables fermions to escape from the Dirac sea is *chiral*, a conceptual framework elaborated in Chapter 5.

We shall provide a vantage point to interpret dark energy as the surplus in vacuum energy from the amount required to maintain the topology of the universe as it expands. The concentration of this surplus is indeed constant since dark energy is not subject to dilution associated with vacuum creation. This fact stands in contrast with the concentration of dark matter that will be shown to be in dynamic equilibrium with detectable matter. Clearly, the dark matter concentration decreases as it gets progressively diluted because the vacuum grows larger over time. The sustainability of the universe ensures the dynamic equilibrium between "darkened" fermions drawn from the Dirac sea and detectable matter, a contribution which maintains the overall mass of dark matter constant while the overall concentration (but not amount) of dark energy is kept unchanged.

At this point we can anticipate that the dynamic equilibrium between dark matter and detectable matter is maintained by a cosmic engine fueled by vacuum energy that sustains the portal to the dark universe while maintaining a constant concentration of dark energy in an ever-expanding universe. This AI–empowered model will be elaborated in Chapter 5.

REFERENCES

1. Weinberg S (2008) *Cosmology*. Oxford University Press, New York, USA.
2. Fisher P (2022) *What Is Dark Matter?* Princeton University Press, Princeton, NJ.
3. Profumo S (2017) *Introduction to Particle Dark Matter*. World Scientific Publishing, Singapore.
4. Susskind L (1995) The World as Hologram. *J Math Phys* 36: 6377–6396.
5. Fernández A (2022) *Topological Dynamics for Metamodel Discovery with Artificial Intelligence*. Chapman & Hall/CRC, Taylor and Francis, London, UK.
6. Wesson PS (1999) *Space–Time–Matter, Modern Kaluza–Klein Theory*. World Scientific, Singapore.
7. Dirac PAM (1982) *The Principles of Quantum Mechanics*, 4th Edition. Clarendon Press, UK.

CHAPTER 2

Quintessential Extension of Space-Time Yielding Mediators between Visible and Dark Sectors

"Spacetime tells matter how to move;
matter tells spacetime how to curve."
JOHN A. WHEELER

This chapter describes topological features of the universe that seem to have been hitherto omitted since they do not impinge in any obvious way on Einstein's local theory of gravity or even on the nonlocality of quantum physics. This topological characterization invites a veritable deconstruction of the Standard Model of particle physics. In turn, the deconstruction is essential to deal with the problem of whether dark matter may be accounted for by extending the Standard Model with incorporation of an extra spatial dimension. After provisionally settling the problem of the universe topology as a quotient topology, this chapter argues for the existence of a compact fourth spatial dimension that stores dark matter. The quest for dark matter becomes contingent on identifying and validating the topology of the universe as a compact multiply

DOI: 10.1201/9781003501534-3

connected space whose quotient space modulo the "dormant" dimension is homeomorphic to $\mathbb{R}^3/\mathbb{Z}^3 \times \mathbb{R}$. Experimental measurements of the spectrum of the gradient temperature field for cosmic microwave background (CMB) radiation led to the conclusion that a dormant circular dimension is consistent with a present-day toroidal space-time with extremely large aspect ratios relative to the dormant coordinate. The topological invariance of the universe since the big bang is thus upheld (it was once compact hence it remains compact), provided the relativistic picture holds since the cosmic birth.

2.1 DARK MATTER STORAGE REQUIRES AN EXTRA DIMENSION

As discussed in the previous chapter, Newton is credited with providing the first model of gravity which he named law of universal gravitation. The predictive power of this law is staggering and thus it constitutes one of the towering achievements of humanity. Yet, when asked about the nature of gravity, Newton cautiously responded that he would not frame hypotheses. Hence, in this sense, Newton may be said not to have conceived a "theory of gravity." Humanity had to wait until the early 20th century to get the first such theory, embodied in Einstein's general relativity [1]. This theory recognizes gravity as an acceleration exerted by a body endowed with mass and regards the fabric of space-time as if it were a membrane, whereby acceleration becomes effectively commensurate with curvature, in turn determined by mass concentration (Figure 2.1). To further see how Einstein got inspired to describe the fabric of space time, let us examine the equation for a vibrating string that gets perturbed from its equilibrium position along a spatial dimension x by an amount ϕ:

$$\frac{\partial^2 \phi}{\partial t^2} - v^2 \frac{\partial^2 \phi}{\partial x^2} = 0. \tag{2.1}$$

The term on the left indicates the second instantaneous variation of the displacement with respect to time, that is, it represents an acceleration due to an exogenous force (Figure 2.1), while the distortion in the string caused by the force exerted is represented by the curvature $\frac{\partial^2 \phi}{\partial x^2}$ of the string. It should be noted that this second spatial derivative is seldom interpreted as the curvature, which in fact it is, just like the first derivative gives the slope

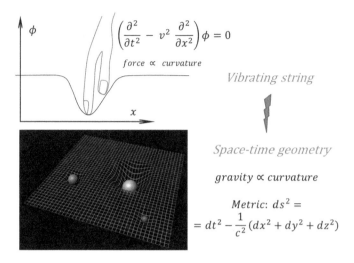

FIGURE 2.1 The equation for the vibrating string where curvature becomes commensurate with the applied force (or acceleration) as inspirational source for Einstein's differential geometry model of general relativity.

of the curve. The proportionality constant is the square of the speed at which the perturbation propagates in time along the x-dimension.

While this fact is typically omitted, it is fairly obvious that the fundamental Equation (2.1) linking curvature and acceleration served as the primary source of inspiration for the differential geometry model of space-time that Einstein adopted in general relativity, his classical theory of gravity (Figure 2.1). It suffices to notice that the metric he adopted in his 4D space-time with coordinates t, x, y, z became:

$$ds^2 = dt^2 - \frac{1}{c^2}\left(dx^2 + dy^2 + dz^2\right). \tag{2.2}$$

This differential volume ds^2 of space-time is the one dictated by the relation between curvature and acceleration, except that the velocity v is now the speed of light c. A complementary way to assert the pivotal relation between curvature and gravity at the heart of Einstein's argument requires that we remind ourselves of another classical equation of physics, the Poisson equation. This equation relates the mass density ρ (amount of mass per unit volume) and the gravitational field ϕ "generated" by the mass. Poisson's equation reads:

$$\nabla^2 \phi = 4\pi G \rho \tag{2.3}$$

where $\nabla^2 = \left(\dfrac{\partial^2}{\partial x^2} + \dfrac{\partial^2}{\partial y^2} + \dfrac{\partial^2}{\partial z^2} \right)$ and $G = 6.674 \times 10^{-11}$ Nm²/kg² is Newton's gravitational constant [1]. Thus, integrating Equation (2.3) on a ball of radius r yields Newton gravitational field

$$\phi = -MG/r, \tag{2.4}$$

where M is the mass contained in the ball of radius r. The dimensions in the r.h.s of Equation (2.3) are those of acceleration (m/s²), and hence, we again can justify the relationship between curvature (measured as $\nabla^2\phi$) and acceleration due to the gravitational pull per unit volume, given by $4\pi G\rho$. Thus, the differential geometry of the space-time manifold of general relativity is inspired and amply justified by the classical equations of physics. This observation prompts us to assert that Einstein's theory of gravity is in fact a classical theory [1]. It is clear that we now have a theory of gravity that includes its interaction with light, a feature absent in the Newtonian law of universal gravitation. Einstein provides a differential geometry framework in which light travels along a hypermembrane (a manifold often referred to as "brane") whose geometric fabric is defined by mass distribution.

Yet, it has proven daunting to reconcile Einstein's theory with quantum physics, now known to govern the other three fundamental forces of nature (electromagnetism and the weak and strong nuclear force). Furthermore, gravity is 10^{-38} times weaker than electromagnetism, an extremely irksome fact that points to a massive geometric dilution of gravity on the known dimensions of space-time. These conundra are of course compounded by the mysterious nature of dark matter that postulates the existence of invisible massive particles that do not detectably interact with the known elementary particles identified in the "Standard Model" [2], and yet they provide the "missing gravity" in the detectable portion of the universe [3].

A window of opportunity for further scientific inquiry into these seemingly related problems and others arising thereof is offered by the possibility of incorporating an extra (fifth) dimension to the differential geometry of the space-time manifold. To enable the possibility of storing significant amounts of kinetic energy in an undetectable stationary wave amenable of a quantum mechanical treatment, the extra dimension would be expected to be compact, specifically rolled up in a circle of extremely small radius. It should be noted that the smallest conceivable material dimension at present is that of a quark (see next section), of the order of 10^{-18} m (a length unit named attometer) [4].

If we specifically adopt a circular fifth dimension of radius $r_0 = 0.802 \times 10^{-18}$ m, we find out, through Einstein's relation $E = hf = hc/\lambda$, that it can store stationary waves with extremely large energies:

$$E_{5,n} = \frac{nhc}{r_0}, n\lambda = 2\pi r_0, \quad \hbar = \frac{h}{2\pi}, \quad n = 1, 2, \dots . \tag{2.5}$$

Strikingly, the lowest such energy is $E_{5,1}$ = 246 GeV, which is precisely the vacuum expectation energy of the elementary particle responsible for bestowing mass on the other known particles, the so-called Higgs boson (see next section) [2]. The daunting problem of incorporating the extra dimension arises from the fact that the circular dimension cannot be thought of as independent of the others, at least there does not seem to be any obvious reason for such an assumption. Hence, the preexisting four dimensions should be considered locally cylindrical, or rather helical instead of linear, with symmetries becoming only approximate and vast differences in curvature depending on the stride of the helices. Such a universe will be described subsequently and will be shown to be far more suitable to achieve a unified field theory while explaining the extreme geometric dilution of gravity on the "observable space-time manifold," as well as the origin of dark matter and dark energy.

Such a fifth dimension would be extremely hard to detect with current experimental resources, even at the LHC operating at optimal performance, due to the extremely high energies associated with wavelengths of the order the quark dimension. Yet, highly massive particles may be yielded by storing energy on the fifth dimension. To visualize this, we note that the kinetic energy along the fifth dimension would be undetectable and hence will be regarded as rest mass, in accord with Einstein's relation [2]. Thus, the total rest mass M of a particle on a 5D space-time manifold would be given by the equation

$$M^2 = m^2 + m_5{}^2 = m^2 + \left(\frac{nh}{cr_0}\right)^2, n = 1, 2, \dots . \tag{2.6}$$

Here m denotes the detectable rest mass that we would be capable of measuring from the particle's existence in Einstein's 4D space-time and

$m_5 = p_5/c$ is the rest mass associated with the component p_5 of momentum along the fifth dimension.

If a particle only stores stationary kinetic energy in the fifth dimension and has no observable rest mass, it is likely to be invisible, yet it could be more massive than any of the known particles. For instance an "ur-Higgs boson" would have the mass corresponding to the vacuum expectation energy of the Higgs boson, at

$$M = M_{u-H} = \frac{\hbar}{cr_0} = \frac{246 GeV}{c^2}. \tag{2.7}$$

This is almost twice the detected mass of the Higgs boson, known to be 124.97 GeV/c². The ur-Higgs and its relatives with masses $M_n = \frac{nh}{cr_0}, n = 2,3,\ldots$ may be regarded as the first particles created after the big bang. We may estimate the lifetime at which such particles were created from Heisenberg's uncertainty relation discussed in Chapter 1:

$$\Delta E \times \Delta t \sim \frac{\hbar}{2} \tag{2.8}$$

Since $\hbar = 6.582 \times 10^{-16}$eVs, we may estimate the time of formation of an ur-Higgs boson ($\Delta E = 246 \times 10^9$eV) at $t = t_{u-H} \approx 1.34 \times 10^{-27}$s. More massive relatives of the ur-Higgs are expected to emerge at shorter times estimated at $t = \frac{t_{u-H}}{n}$.

As we discuss massive relatives of the ur-Higgs we are prompted to ask: How large can n be? The answer is straightforward, as the so-called Planck mass, $M_P = \sqrt{\frac{\hbar c}{G}} = 1.22 \times 10^{19}$ GeV/c², sets an upper bound to the mass, so that if $M_n = \frac{nh}{cr_0} \geq M_P$, then the particle becomes a black hole. Thus, the first particle formed in the universe had a mass M_n^* corresponding to the largest integer $n = n^*$ such that $n^* \leq \frac{M_P}{M_{u-H}} \approx 4.95 \times 10^{16}$. This supermassive particle formed at a time estimated at $t^* = \frac{t_{u-H}}{n^*} \sim 10^{-43}$s, during the so-called "Planck epoch" of the universe.

Thus, a picture of the very early universe for the period 10^{-43}s $\leq t \leq 10^{-27}$s may be obtained by assessing the energy stored in the stationary de

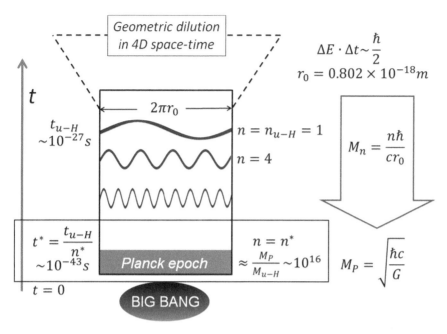

FIGURE 2.2 Elementary particles in an early universe covering the period $10^{-43}\text{s} \leq t \leq 10^{-27}\text{s}$ after the big bang obtained by storing energy as stationary waves that span the fifth dormant dimension.

Broglie wave that spans the fifth dimension, as schematically depicted in Figure 2.2.

2.2 TOPOLOGY STEERS THE INCORPORATION OF AN EXTRA DIMENSION

We have thus seen that an extra spatial dimension is invoked to address a specific need requiring an extension of a preexisting model. This intellectual procedure is fairly common and can be found in diverse areas of knowledge. For example, we may ask why living organisms cannot exist as 2-dimensional objects, and more generally, why life requires at least three dimensions. The answer is simple: Because 2-dimensional organisms could not metabolize, since no digestive tract would be possible in two dimensions (Figure 2.3). Furthermore, we have tangible evidence of the existence of extra spatial dimensions in the so-called quasi crystals, which may be thought of as projections in two dimensions of 5-dimensional crystalline structures (Figure 2.4).

The most difficult issue we stumble upon when postulating a compact fourth spatial dimension to store dark matter is that the global topology of

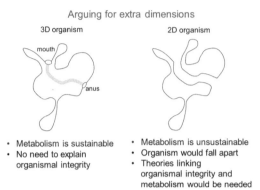

FIGURE 2.3 Extra dimensions required to define a living organism vis-à-vis the impossibility of a 2-dimensional organism capable of metabolic activity.

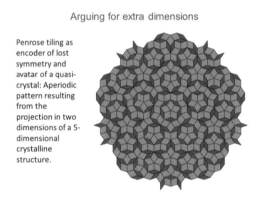

FIGURE 2.4 Penrose tiling as avatar of a quasicrystal.

the universe must allow for such possibility. In the past, people have tried to incorporate a curled-up dimension into an Euclidean space, an idea that is topologically incompatible yet, of course *a priori* possible. Not only such attempts fail to generate correct extensions of general relativity, as discussed in Chapter 1, but they arose from the unappealing and unwarranted assumption that one dimension would be compact, while the others are infinite.

Incorporating a compact dimension is a nuanced matter that pivotally depends on the topology of the universe. This was clearly not a matter of concern for Einstein because his general relativity is essentially governed by differential equations. Since differential equations describe a local situation, the global topology of space-time plays no obvious role. Furthermore, relativity has no concern for boundary conditions, as the universe is assumed to be infinite. In our topology, the universe is compact and multiply connected but there are no boundary conditions, simply because there is no boundary.

On the other hand, quantum mechanics defines the fabric of space, as determined by vacuum entanglement, but topology is not factored into the theory in any obvious way, even if entanglement is thought to be the culprit for the nonlocality of quantum phenomena. However, topology suddenly becomes extremely relevant as one tries to reconcile general relativity with quantum physics, as it is the case with quantum gravity, as shown in Chapter 5.

If we assume topological invariance of space throughout the universe evolution starting at the big bang event, then space cannot be infinite and flat, and therefore Euclidean. Provided we accept the big bang scenario, space must be compact because it once was compact with certainty. But if space is compact, then it must be multiply connected. In addition, space cannot have a boundary because that would imply an interface with nothingness, and it is physically and metaphysically impossible to interface with nothingness: An interface presupposes two media. These considerations introduce several constraints on what the topology of the universe must be. Summarizing, space must be compact, locally flat, orientable, and lacking boundary. This leaves us with essentially one option, and it is a multiply connected 3-dimensional manifold: The 3-torus or Cartesian product $S^1 \times S^1 \times S^1$. Hence, the incorporation of a fourth circular dimension is compatible with the topology of the space-time manifold.

There is a fundamental difference in the way the extra spatial dimension, the curled dormant dimension, is incorporated in the treatment presented in this book when compared with the early attempts, especially the so-called Kaluza-Klein model [2]. These researchers incorporated the extra dimension *directly* to the Euclidean space, creating the absurdity of having one compact and three infinite spatial dimensions in space-time with no justification. This construction proved to be untenable from every perspective, a mere intellectual curiosity.

As we have shown, the spatial cross-sections of space-time cannot be regarded as isomorphic to $\mathbb{R}^3$ but rather to the 3-dimensional torus $T^3 \approx \mathbb{R}^3/\mathbb{Z}^3$, that is, a quotient space of classes modulo all the spatial translations of the unit cell indexed by three integers (Figure 2.5). This implies that the space is multiply connected with all spatial dimensions being compact, hence allowing for the incorporation of an extra circular dimension ($S^1 \approx \mathbb{R}/\mathbb{Z}$) without the conceptual violence of the Kaluza-Klein

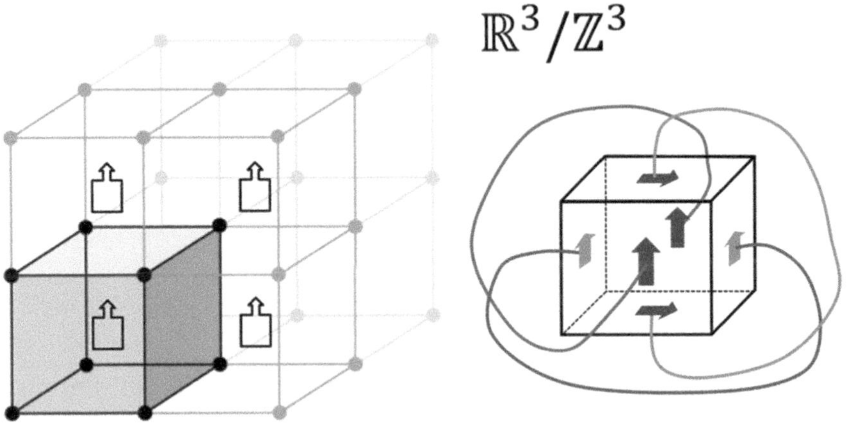

$$\mathbb{R}^3/\mathbb{Z}^3$$

FIGURE 2.5 Generic spatial cross section of space-time with compact coordinates, as required by the topological invariance of the universe dictated by the tenets of general relativity. The compact space is isomorphic to the 3-dimensional torus in accord with the topological equivalence $\mathbb{R}^3/\mathbb{Z}^3 \approx T^3$. The latter manifold is obtained by identifying opposite faces of the solid cube in three dimensions, hence featuring a wormhole upon construction with dimensional aspect ratios equal to 1 (Chapter 5).

models. A hybrid $\mathbb{R}^3 \times \mathbb{R}/\mathbb{Z}$ of three infinite and one compact spatial dimension would make no sense and is in fact not compatible with the tenets of general relativity, assuming this theory is upheld since the inception of the universe, as topological change is forbidden.

At a very fundamental level, *the Poincaré group, assumed to be the group of isometries of space-time over which the whole edifice of the Standard Model is built, does not really contain the proper symmetries but only approximate symmetries, or rather pseudo-symmetries, since the underlying spatial cross sections of space-time are not isomorphic to $\mathbb{R}^3$ but to $\mathbb{R}^3/\mathbb{Z}^3$.*

On the other hand, a spatially compact universe implies the existence of a primeval wormhole, as inferred from Figure 2.5 and demonstrated in Chapter 5. This can be visualized in a 2-dimensional cross section T^2 with dimensional aspect ratio 1. As shown in Chapter 5, the sustainability of this wormhole is concomitant with the universe expansion. On the other hand, the very notion of universe expansion becomes meaningless in the noncompact space $\mathbb{R}^3$, while it makes good sense to talk about expansion of the unit cell in the quotient space $\mathbb{R}^3/\mathbb{Z}^3$. This is yet another reason to assume that the spatial cross sections of the relativistic space-time are actually isomorphic to $\mathbb{R}^3/\mathbb{Z}^3$, and not to $\mathbb{R}^3$, as assumed in the Minkowski space-time endowed with the isometries of the Poincaré group.

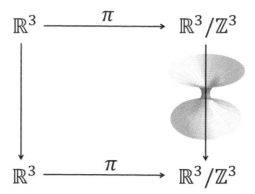

FIGURE 2.6 Commutative scheme adopted by artificial intelligence to circumvent the artefactual wormhole phenomenology associated with the universe compactness by lifting it at the level of universal covering ($\mathbb{R}^3$, π) of the toroidal spatial cross-sections ($\mathbb{R}^3/\mathbb{Z}^3$) of space-time.

On the other hand, AI adopts the commutative scheme presented in Figure 2.6, where the autoencoder defines the lifting of the Standard Model at the level of the universal covering with canonical projection π: $\mathbb{R}^3 \to \mathbb{R}^3/\mathbb{Z}^3$, so that the diagram presented in Figure 2.6 is commutative. This lifting has considerable advantages since it circumvents the observed universe expansion with the concomitant primeval wormhole as artefactual of "Plato's cave physics," where the observer is placed. Even dark energy, shown in Chapter 5 to be the residual vacuum energy not invested in sustaining the primeval wormhole, becomes artefactual of Plato's cave phenomenology in $\mathbb{R}^3/\mathbb{Z}^3$, the only one accessible to the quantum observer.

2.3 ELEMENTARY PARTICLES AS EXCITATIONS OF QUANTUM FIELDS

We now know that atoms are not the indivisible constituents of matter that Democritus and other pre-Socratic philosophers imagined. Atoms are made of even smaller bits, called elementary particles. Yet, popular accounts notwithstanding, there is nothing "corpuscular" about elementary particles when regarded in the current sense of the word. Major modeling efforts that followed after the advent of general relativity and quantum mechanics revealed that the true nature of elementary particles cannot be properly captured in the corpuscular representation. Rather, elementary particles constitute excitations or warps in fields – one for each particle type – that establish a correspondence between each point of space with a scalar or vector value [2]. This is the spirit of the quantum field theory (QFT) that spearheaded current modeling efforts in elementary particle physics.

We are familiar with such vector or scalar fields in physics. For example, a temperature (T) field is a scalar field that assigns a T-value to each point in space. A gradient in T (if the field is smooth or differentiable) steers heat convection, that is, energy transference from the region with higher heat content (at higher T) to the region of lower heat content (at lower T). Thus, we may say that energy (E) is the magnitude associated with the T-field, as reflected in Boltzmann's equation $E = kT$, where k is Boltzmann's constant. Likewise, a dynamical system may be defined by a velocity field, a vector field that determines the trajectories as curves that are tangent at each point to the vector assigned by the field. Similarly to the E-T duality, elementary particles have been defined in QFT vis-à-vis fields that represent them. An elementary particle becomes a local concentration of energy or its equivalent mass, as dictated by Einstein's relation $E = mc^2$ derived in Chapter 1. The particle field itself is viewed as a map $\phi: W \to V$, where W is the 4-dimensional space-time of special relativity and V is either $\mathbb{R}$, $\mathbb{C}$, the sets of real or complex numbers, or $\mathbb{R}^n$ ($\mathbb{C}^n$), $n > 1$, depending on whether ϕ is a scalar or a vector field, respectively. Just like E may be viewed as an excitation of its dual T field, the picture where particles are regarded as excitations of their respective fields is a major tenet of QFT and the modeling effort resulting thereof is known as the Standard Model (SM) of particle physics [2].

In accord with current thinking, particles are best characterized as local modes of storing energy/mass, that is, localized field excitations defined by independent representations of the symmetry group of space-time [2]. These representations have generators that are parametrized by specific attributes of particles, including charge, mass, spin (an intrinsic angular momentum vis-à-vis an internal axis of rotation), spin degrees of freedom (projection of the spin vector along the direction of particle motion), and so on. Yet, this manner of particle classification tells us very little unless we can define interactions between particles based on their intrinsic attributes, assessing how particles communicate and transform into one another. The nature of these interactions is delineated in the next section.

Now let us briefly explore how the concept of particle field came about. By attributing the energy of the photon to its frequency of oscillation as in the equation $E = hf = hc/\lambda$, Einstein was the first to associate a particle with a wave in his description of the photoelectric effect. Einstein, and Planck before him, reasoned that the frequency was simply a surrogate for kinetic energy, as it is an indicator of the number of cycles per unit time. This was followed by de Broglie's representation of the electron orbiting the atomic nucleus as a stationary wave with a wavelength that satisfies the relation $n\lambda = 2\pi r$, where n is a positive integer value and r is the orbit radius. This "quantization" of the wavelength ensures constructive (reinforcing) phase interference, as depicted in Figure 2.7. It also captures the discreteness of the experimentally obtained electron absorption spectrum, as the only allowed transitions for the electron are the result of absorption of photons carrying energies given by the differences between the n-indexed electron energies $E_n = n\hbar c/r$, with $\hbar = h/2\pi$. In fact, de Broglie went one step beyond, as he associated any particle with momentum p to a wavelength λ according to the relation $p = h/\lambda$. We can see in these early modeling efforts the need to go beyond a mere corpuscular description of particles, a conceptual framework later adopted by QFT in its construction of the SM [2].

The SM has defined, characterized, and organized all of the known elementary particles in much the same way that the periodic table systematically categorized or classified the types of atoms known as elements. The predictive value of the SM has been staggering, notwithstanding the fact that it does not fully incorporate gravity, while being able to encompass the other three quantum forces of nature in a unified way.

At the turn of the 20th century, it was believed that there were only three fundamental particles in nature: Protons, neutrons, and electrons, with the first two making up the atomic nucleus, while the electrons revolved around it, as in a mini planetary system. But by the 1960s physicists started smashing these particles together and were able to show

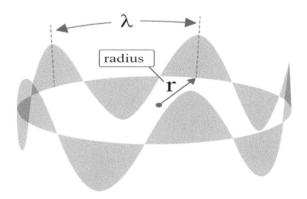

FIGURE 2.7 A de Broglie stationary wave of wavelength λ in a circle of radius r.

unambiguously that protons and neutrons were made up of even smaller particles. Many particles were discovered in a relatively short period of time and for a while it was not possible to make much sense of the "particle zoo."

Yet, by 1970, mainly due to the monumental "bottom-up" modeling efforts based on QFT, elementary particles were shown to fall in two main categories: Fermions and bosons, with fermions making up matter while bosons, or more precisely "gauge bosons," transmit forces to which fermions respond or react to [2]. Fermions are divided into two kinds of particles, depending on the forces they respond to: Quarks and leptons. Particles within the three types – quarks, leptons, and bosons – are symmetry related to each other. This means that a coordinate transformation associated with a space-time symmetry operation transforms one particle into another of the same family. It was established that particles communicate with one another via four forces: Electromagnetism, strong nuclear force (SNF), weak nuclear force (WNF), and gravity. The SM describes the first three, while gravity does not yet feature satisfactorily in the current version of the SM, as already anticipated.

Different particles communicate and interact through different forces. For example, only the quarks relate to the gluon, its natural gauge boson and carrier of the SNF, while electrons relate to the photon, the gauge boson that carries the electromagnetic force. In addition, electrons can also communicate via the W boson and Z boson, the carriers of the weak nuclear force. So electromagnetism is the force that holds electrons in the atom, while the SNF communicated by gluons keeps the nuclei from breaking apart. Meanwhile, the WNF communicated by Z or W bosons assists the radioactive decay of nuclei, whereby, for example, the quark/gluon assemblage of a neutron composite becomes that of a proton upon interaction with a W boson with emission of electromagnetic particles.

Quarks come in six "flavors" that in turn come in pairs to make three generations. These are "up" and "down" (first generation), "charmed" and "strange" (second generation), and "top" and "bottom" (third generation). The up and down quarks are relevant to atomic physics because they make protons and neutrons, while the others make "exotic" matter, too unstable to form atoms and usually lasting a fraction of a second before decaying.

There are six leptons. The best known is the electron, a fundamental particle with a negative charge. The muon (second generation) and tau (third generation) particles are heavier versions of the electron. They also have negative electric charge, but they are too unstable to feature in ordinary matter. And each of these particles has a corresponding neutrino (Italian for "small neutron"), a spin carrier with no charge and almost no mass. Neutrinos interact via the WNF, which is known to cross talk to electromagnetism, as in the radioactive decay events mentioned previously.

Quarks and leptons have twin particles of antimatter known as antiparticles. Antimatter differs from matter only in that it has the opposite charge. For example, the electron has an antimatter counterpart, the positron, with the same mass but positive charge. When a particle meets its antiparticle, they both annihilate in a burst of energy carried by a boson.

We shall now characterize the Higgs boson, a particle endowed with a field ϕ that may be represented as a doublet of complex numbers: $\phi = (\phi_1, \phi_2)$, with $\phi_j = \mathrm{Re}\,\phi_j + i Im\phi_j$, $j = 1, 2$. This field, empirically determined in a brilliant modeling effort, is peculiar in that its potential energy does not reach its minimum at zero field but a value of ~246 GeV (1 eV = 1.602×10^{-19} J, 1 GeV = 1.602×10^{-10} J) known as the vacuum expectation value (v), hence implying a symmetry breaking (Figure 2.8). The symmetry breaking arises because different changes in potential energy around the zero-point energy are obtained depending on the direction of variation of the field. For example if we write the singlet field in radial coordinates as $\phi = \rho e^{i\theta}$, then the zero-point energy is the circle $\rho = v$ and radial variations $\rho = v \pm \delta h$ yield variations in the potential energy (dashed line in Figure 2.8), while angular variations along the circle $\rho = v$ yield no change in potential energy, which remains at its zero value. This symmetry breaking turns out to be of paramount importance in determining the role of the Higgs boson vis-à-vis its interactions with other particles. Thus, just like other gauge

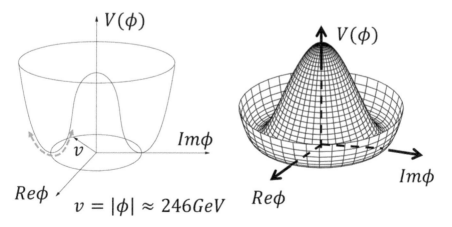

FIGURE 2.8 Scalar field for the Higgs boson.

bosons are force carriers, communicating different types of forces associated with particle interactions, the Higgs boson may be interpreted as bestowing mass upon other particles.

To interpret this process of mass endowment as well as the communication of fundamental forces by the other gauge bosons it is necessary to resort to the mathematical arsenal of particle physics [2]. This arsenal is necessary to describe the "geodesic flow," that is, the lines of least action followed by the fields that underlie particle physics. As said, a basic tenet of QFT indicates that a particle is represented as a local excitation of its respective field, representing a local concentration of energy or, equivalently, of mass. In QFT, a particle scalar field $\phi = \phi(\{x_\mu\}_\mu)$ in space-time is determined by the particle Lagrangian $\mathcal{L} = \mathcal{L}\left(\phi, \{\partial_\mu \phi\}_\mu\right)\left(\partial_\mu \equiv \dfrac{\partial}{\partial x^\mu}\right)$ whose definition is inspired by classical mechanics: $\mathcal{L} = \dfrac{1}{2}\left(\partial_\mu \phi\right)^2 - V(\phi)$, with the first term on the r.h.s. representing the kinetic energy (K), and the second, the potential energy. The latter is exemplified in Figure 2.8 for the Higgs field. Einstein convention of summation over repeated indices is followed throughout. Thus, the geodesic flow in space-time defined by the field $\phi = \phi(\{x_\mu\}_\mu)$ minimizes the action $\mathfrak{A} = \displaystyle\int \mathcal{L}\left(\phi, \partial_\mu \phi\right) dx^\mu$ and hence the field satisfies the so-called Euler-Lagrange equations:

$$\partial_\phi \mathcal{L} = \partial_\mu \left[\frac{\partial \mathcal{L}}{\partial(\partial_\mu \phi)} \right]. \tag{2.9}$$

To see how two particles, determined respectively by fields ϕ_1, ϕ_2, interact, or one particle communicates a force to another particle, we need to consider the sum $\mathcal{L}_1\left(\phi_1, \{\partial_\mu \phi_1\}_\mu\right) + \mathcal{L}_2\left(\phi_2, \{\partial_\mu \phi_2\}_\mu\right)$ of their respective Lagrangians. An appropriate coordinate transformation $x^\mu \to x'\nu$ and suitable transference or swapping of terms yields the sum: $\mathcal{L}'_1\left(\phi'_1, \{\partial_\nu \phi'_1\}_\nu\right) + \mathcal{L}'_2\left(\phi'_2, \{\partial_\nu \phi'_2\}_\nu\right)$, satisfying the action equality

$$\int \left[\mathcal{L}_1\left(\phi_1, \{\partial_\mu \phi_1\}_\mu\right) + \mathcal{L}_2\left(\phi_2, \{\partial_\mu \phi_2\}_\mu\right) \right] dx^\mu$$

$$= \int \left[\mathcal{L}'_1\left(\phi'_1, \{\partial_\nu \phi'_1\}_\nu\right) + \mathcal{L}'_2\left(\phi'_2, \{\partial_\nu \phi'_2\}_\nu\right) \right] dx'^\nu. \tag{2.10}$$

This implies that the original interactive particle pair defined by the fields ϕ_1, ϕ_2, has transformed through mathematical manipulation (including only term reorganization and change of variables) into new particles inter-preted as being defined by the fields ϕ'_1, ϕ'_2. These destiny fields are asso-ciated with the "destiny Lagrangians" $\mathcal{L}'_1$, $\mathcal{L}'_2$ and may correspond to any number of particles, elementary or composite.

For example, let us consider the interaction of the Higgs field $\phi_1 = \phi_H$ (Figure 2.8) with a second field ϕ_2 corresponding to a particle that may be the original massless version of the W or Z boson. The kinetic energy K_H for the Higgs field may be written in terms of the covariant coordinates $D_\mu = (\partial_\mu - i\phi_2)$ that respond to the field of the other particle. Upon the coordinate changes $\phi_H \rightarrow (\rho, \theta)$ and substituting $\rho = v + \xi$ around the v-value ring of ϕ_H (Figure 2.8), the kinetic energy K_H reads (asterisk here denotes complex conjugate):

$$K_H = \frac{1}{2}\left(D_\mu\phi_H\right)\left(D_\mu\phi_H\right)^* = \frac{1}{2}\left(\partial_\mu\xi\right)^2 + \frac{1}{2}\phi_2^2\left(v+\xi\right)^2. \qquad (2.11)$$

This implies that the precursor to the gauge boson has now gained the potential energy term estimated as $\frac{1}{2}\phi_2^2 v^2 : \mathcal{L}'_2 \approx \mathcal{L}_2 + \frac{1}{2}\phi_2^2 v^2$ which implies that the mass bestowed by the Higgs field corresponds to radial (i.e., along the ξ–coordinate) excitations of its field, as depicted in Figure 2.8.

2.4 BIG BANG SCENARIO IN A RELATIVISTIC UNIVERSE: A RUSSELL-TYPE PARADOX

The forthcoming "metaphysical" discussion is inevitable, since we shall discuss the big bang scenario in connection with the nature and origin of dark energy and dark matter. Space-time, a foundational concept in relativity, incorporates time as an inherent dimension, so we may assert that space-time per se, and hence the universe, cannot evolve. *Prima facie* this statement may seem disconcerting, but considering the evolution of space-time introduces a paradox of the Russell type. The point is that the phrase "evolution of space-time" is meaningless as no entity where time is immanent in its being can actually evolve. Since time is inherent to space-time and inseparable from the spatial dimensions, how could space-time change over time? All time change in space-time is already contained in space-time (Figure 2.9).

What we presumably mean by "evolution of the universe" is actually something quite different: Consider "formally" a fixed-t cross section of space-time and define evolution of the universe as the sequence of t-cross

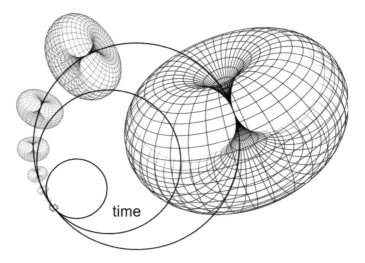

FIGURE 2.9 Space-time subsumes its own evolution, hence the phrase "space-time evolution" is rigorously meaningless and entails a Russell-type paradox. A torus topology of space cross sections is justifiably adopted.

sections as the "parameter" t is varied. In other words, we consider the time evolution of space. While this seems *a priori* a well-defined object, it is immanent in space-time, so it cannot be equated, even by abuse of language, with the latter. Furthermore, space-time does not allow for the dissociation of time from spatial dimensions. This is a pillar of relativity. There is no external clock. The dimensions are not "separable." For example, if we query about the big-bang scenario vis-à-vis the universe evolution, we are rapidly trapped into a paradox, because the extreme mass concentration at the "origin" prevents time from flowing, hence *the primeval point is* a priori *a state of the universe from which the universe can never evolve*. This is an extreme case of the time dilation experienced by entities crossing the event horizon of a black hole. In fact, a trajectory leading out of a black hole would entail time reversal (Figure 2.10), so time cannot be "externalized" at the big bang, let alone flow to describe the origin of the universe.

Space-time postulates interdependence between spatial dimensions and time. As such, the dimensions are not separable or cannot be regarded as independent. Thus, fixed time cross sections may constitute a "formally" acceptable concept but the time evolution of such cross sections is problematic especially given that the magnitude of time dilations is strongly dependent on space. This discussion brings us back to the big bang scenario. If space is concentrated in a geometric point and time is

space-time cross section

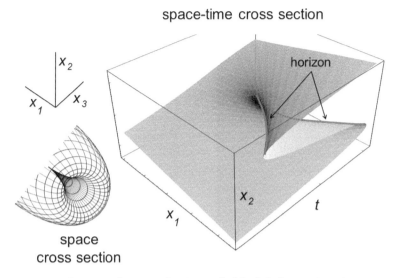

space
cross section

FIGURE 2.10 Crossing the event horizon of a black hole requires time reversal, as depicted in the space-time model of the singularity.

consequently dilated to infinity, how could the universe be said to have ever evolved in the big bang scenario? Furthermore, what is meant by evolution of the universe, when space-time evolution, understood in any acceptable way, is subsumed in space-time itself (Figure 2.9)? It appears that the big bang scenario is ill posed, and even the phrase "universe evolution" is ill defined.

Yet, the discussion on universe evolution and big bang is alive and ongoing and has borne fruits, especially in the realm of quantum cosmology [1]. This makes it necessary to come up with an operational definition of "evolution of the universe" or at least, of "evolution of space" that is not conceptually violent and can even reconcile the inseparability of space and time that is inherent to the relativistic perspective. A possible *ansatz* put forth in this book is as follows: Assume a time dilation defined by $t_v \rightarrow t = t_v/\vartheta(x)$, scaling time flowing in an inertial vacuum (t_v) according to a spatially dependent factor $\vartheta(x)$, inherently dependent on the local curvature tensor and commensurate with its norm. We define *extrinsic evolution of space* as the evolution of the spatial t-cross section of space-time parametrized by rescaling time to make it an extrinsic parameter dissociated from the intrinsic time $\left(t = \dfrac{t_v}{\vartheta(x)} \right)$ that is fixed as constant in determining the cross section. In other words, we consider the "extrinsic evolution" of the t-slice of space-time: $\mathcal{X}\left(t_{ext}\right)=\left\{x=x\left(t\right)\right\}_{x}$ re-parametrized for each point in space as dependent on "extrinsic time": $t_{ext} = t \times \vartheta(x(t))$.

For relativistic singularities such as the black hole, the scaling of time dilation does not hold. This is because to escape from a black hole horizon the speed of the moving body should exceed the speed of light, implying that time should move backwards. This can be readily visualized in a cross section of the space-time manifold at the black hole, incorporating two spatial dimensions and time (Figure 2.10). In general, following Hendrik Lorentz [1], we can make the time scaling dependent on the speed of the moving body in accord with the so-called Lorentz transformation: $(v) = \sqrt{1 - \dfrac{v^2}{c^2}}$, however, the Lorentz formula yields an imaginary factor for speeds exceeding the speed of light ($v > c$), which is meaningless in this context and also incorrect because it does not reflect the actual reversal of time associated with the escape from the black hole. The time reversal implies that the escaping observer has actually traveled into the future as it transposed the event horizon of the black hole, and therefore its escape from the black hole would entail traveling into the past, as depicted in Figure 2.9. Since time reversal is a relativistic impossibility, there is no escape from the black hole.

These arguments lead us to formulate the following assertion:

Proposition 2.1

Only one of the following statements is true:

a) The relativistic tenets hold for the big bang and then the big bang scenario does not represent the origin of space-time because space-time subsumes its own evolution and hence it does not have an origin,

b) the relativistic tenets of space-time do not hold in the earliest universe because time and space are decoupled, hence separable, or

c) the relativistic tenets hold for the big bang and then it is impossible to trace the origin of the universe starting at a space-time singularity because no extrinsic flowing time exists or can be defined at its spatial locus. ▪

This proposition leaves AI with fewer resources to tackle the problem of the nature of dark matter and dark energy, as AI builds up theory only if it finds leads to a cogent picture through conjecture framing. As hinted by the results previously described in this chapter, the study of the origin of singularities in space-time is essential to make progress in understanding the nature of dark matter and, as we shall see in Chapter 5, the same holds for dark energy. Yet, it appears that the very phrase "origin of singularity" is troublesome and alludes to an ill-posed problem. The only alternative that has led to further progress is statement b), which is as likely or unlikely to be true as the other two, but the only one pregnant with possibilities for theoretical progress, as we shall demonstrate. Thus, an operational definition of *extrinsic evolution of space* based on the assumption that statement b) is valid will be adopted throughout in the rest of the book and by abuse of notation it will be referred to as "evolution of the universe" or evolution of space-time with the caveat that the clock can be placed outside of the object of study and that it is actually ticking.

2.5 THE 4-TORUS TOPOLOGICAL MODEL OF THE UNIVERSE ADMITS A COMPACT EXTRA DIMENSION

As earlier described, Einstein's theory of general relativity is formulated in terms of differential geometry and its laws are therefore cast as differential equations. It is therefore a local theory. It tells us nothing about the overall shape or topology of the universe while it informs on its geometry [5]. In principle, more than one topology can fit the postulated geometric flatness, which at any rate can only be regarded as an approximation [1, 5]. There is one thing that general relativity does tell us in regard to topology: *As the universe evolved in time since the big bang its topology could not change* [1]. This is fairly obvious, since a topological change would imply cutting and pasting space-time, an inadmissible operation. This is so because it would entail a non-homeomorphic distortion of the differential geometry fabric, unless a new force beyond the four established forces [2] could be invoked.

In general relativity we assume that the intrinsic geometry of the universe 4D-manifold has infinite curvature radius (flat geometry). However, if we soften the assertion to indicate "immeasurably large radius," we may find alternative topologies for compact (yet enormous) manifolds that would fit the postulated approximate flatness. From the previous discussion, it seemed that identifying dark matter and dark energy would require a reverse engineering of the SM to incorporate all dimensions in an early universe together with the evolution of their aspect ratios. Yet, this reverse engineering depends pivotally on the topology of the universe [6], which we know must be invariant in time. This implies that the present-day universe cannot be infinite for the simple reason that the big bang singularity and the early universe were compact. *Hence, our universe must be compact, however enormous.*

A quasi-flat compact manifold must necessarily be multiply connected and hence, the incorporation of a fifth circular coordinate yields one and only one manifold as the only alternative: A 5-dimensional torus, that is, a Cartesian product of five circles with huge aspect ratio between the four circular coordinates conforming the locally quasi-flat 4-dimensional space-time and the extra fifth spatial coordinate with a quark-size (i.e., LHC–undetectable) radius. We regard the standard 4-dimensional space-time in general relativity as a 4-torus locally homeomorphic to the Euclidean space. This is a latent compact manifold representing a quotient space denoted $W/\sim$ [6], so that two points in a higher dimensional space W, the 5-dimensional torus with a circular extra dimension, would be equivalent modulo "$\sim$" if they projected onto the same point in the latent manifold. The task ahead then becomes to elucidate how dark matter would fit in this scheme as a gravity-carrying particle enshrined in the compact fifth dimension, existing in W but not in its quotient space $W/\sim$. In this way, dark matter would not interact with the SM in 4D space-time, except via gravity.

The postulated toroidal topology of the universe has found recent experimental validation in the examination of the CMB (Cosmic Microwave Background), the cosmic relic of the big bang (Chapter 1). If space were infinite (hence flat and simply connected), perturbations in the temperature gradient field of the CMB radiation would exist on all scales. If space is finite, then there would be wavelengths missing that are larger than the size of space itself. Maps of the CMB perturbation spectrum made with probes like NASA's WMAP and the Planck probe from the European Space Agency have shown striking amounts of missing perturbations at large scales. The properties of the observed fluctuations of the CMB show a "missing power" on scales beyond the size of the universe [7]. That would imply that our universe is multiply connected and finite. The spectrum of the CMB is compatible with a 3-dimensional torus topology for the spatial components of the universe. Thus, universe models with spatially multiply connected topology contain a discrete spectrum of the Laplacian with specific wavelength cutoff, as observed in the CMB. Furthermore, the 3-dimensional torus model possesses a two-point correlation function that fits CMB maps obtained with the Planck probe [7].

Thus, the toroidal topology enables the incorporation of the extra curled-up dimension as an undeveloped feature in the compact universe evolution. In principle, nothing precludes the presence of more than one such dormant dimensions. The dormant dimension cannot be probed with current means as it spans the quark attometer scale, storing energy in the order of 246 GeV. On the other hand, the flat, simply connected universe that we all intuitively have come to grips with would actually be incompatible with the evolution from a compact manifold as implied by the big bang scenario simply because such evolution would entail a change in topology. Such a transition from compact multiply connected to flat simply connected cannot be smooth and would be forbidden by general relativity.

Within the compact multiply connected model of today's universe, four of the circular dimensions have grown so much that the curvature radius is for all practical purposes infinite, and the geometry is locally that of a Euclidean space [1]. Thus, the dormant fifth coordinate combines with the quasi-Euclidean standard coordinates in a local cylinder of quark-size radius, where the standard coordinate constitutes the cylinder axis. In

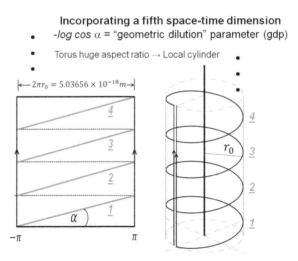

Incorporating a fifth space-time dimension
- *-log cos* α = "geometric dilution" parameter (gdp)
- Torus huge aspect ratio → Local cylinder

$\leftarrow 2\pi r_0 = 5.03656 \times 10^{-18} m \rightarrow$

FIGURE 2.11 Incorporating the circular fifth dormant dimension and combining it with generic locally flat dimensions of the 4-dimensional space-time through a geometric dilution parameter determining the helical stride.

essence, the torus with huge aspect ratio becomes locally a cylinder (Figure 2.11). Standard space-time coordinates get mixed with the dormant coordinate as determined by a pitch angle α defining the helix stride, so that a geometric dilution (v) for the energy stored in the dormant coordinate may be defined as $v = - \log \cos \alpha$. Thus, the geometric dilution becomes infinite when there is no projection along the dormant coordinate and zero, when there is no projection onto the standard coordinates.

2.6 TOPOLOGICAL METAMODEL OF THE UNIVERSE IN THE AI QUEST FOR DARK MATTER

Current knowledge of dark matter is sketchy at best and mostly conjectural. What we actually know or can judiciously conjecture is best summarized as follows:

- Dark matter is invisible, massive, "cold" (speed far lower than the speed of light, $v \ll c$), mainly formed early on, interacts with the SM only via gravity, and does not decay easily.

- It is probably not an extension of the SM, as no particle derived from the latter fits the characteristics of dark matter [3].

- It may be stored in a compact fourth spatial dimension.

- In advocating for storing dark matter in an extra dimension, we note that the quark scale ($q = 0.802 \times 10^{-18}$ m) is the smallest – undetectable – material scale known, and that a stationary wave along a circular fourth spatial dimension with radius q stores an energy $E = hc/2pq = 246$ GeV, the vacuum expectation value of the Higgs boson.

- We have a validated topological model of the compact universe to incorporate such a dimension and trace the origin of dark matter to the early universe.

The core involves the identification of dark matter as stored in the 5-dimensional "quintessential" space W, a 5-dimensional torus projecting onto a 4-dimensional "latent" manifold, $\Omega = W/\sim$ that is locally flat, meaning it is locally homeomorphic to a Euclidean space. Thus, Einstein's 4-dimensional space-time Ω is regarded as a quotient space for the space that we intend to prove stores dark matter. This problem requires the deployment of AI, as it requires the extension of the SM to incorporate a fifth dimension, so far defined in the latent quotient space

$\Omega = W/\sim$. This task, undertaken in Chapter 4, entails the lifting of each pairwise particle interaction within Ω, represented by the transformation:

$$\mathcal{L}_1\left(\phi_1, \{\partial_\mu \phi_1\}_\mu\right) + \mathcal{L}_2\left(\phi_2, \{\partial_\mu \phi_2\}_\mu\right) \rightarrow \mathcal{L}'_1\left(\phi'_1, \{\partial_\nu \phi'_1\}_\nu\right) + \mathcal{L}'_2\left(\phi'_2, \{\partial_\nu \phi'_2\}_\nu\right),$$

to the level of W:

$$\widetilde{\mathcal{L}}_1\left(\widetilde{\phi}_1, \{\partial_\mu \widetilde{\phi}_1\}_{\mu=1,\dots,5}\right) + \widetilde{\mathcal{L}}_2\left(\widetilde{\phi}_2, \{\partial_\mu \widetilde{\phi}_2\}_{\mu=1,\dots,5}\right) \rightarrow \widetilde{\mathcal{L}}'_1\left(\widetilde{\phi}'_1, \{\partial_\nu \widetilde{\phi}'_1\}_{\nu=1,\dots,5}\right)$$

$$+ \widetilde{\mathcal{L}}'_2\left(\widetilde{\phi}'_2, \{\partial_\nu \widetilde{\phi}'_2\}_{\nu=1,\dots,5}\right) \tag{2.12}$$

where $\widetilde{\mathcal{L}}_n, \widetilde{\phi}_n$ denote respectively the lifting of Langrangian $\mathcal{L}_n$ and particle field ϕ_n, with n indicating particle types in the SM. As indicated earlier, the primes denote destiny particles arising from the interaction.

A lifting of the SM is valid if and only if the diagram in Figure 2.12 is commutative. This property indicates that a canonical projection $\pi : W \rightarrow \Omega$ followed by interaction within the latent space Ω yields a destiny state that is the same as that obtained when the interaction is computed directly in the quintessential space W and the resulting destiny state is subsequently projected onto the latent space Ω. In other words, the following equation must hold for every particle pair $\left(X_J\left(\mathcal{L}_J, \phi_J\right), J = 1,2\right)$ in the SM:

$$X'_j\left(\mathcal{L}'_J, \phi'_J\right) = \pi \widetilde{X}'_j\left(\widetilde{\mathcal{L}}'_j, \widetilde{\phi}'_j\right). \tag{2.13}$$

The lifting $\widetilde{\phi} : W \rightarrow \mathbb{C}^n$ of a particle field $\phi: W/\sim \rightarrow \mathbb{C}^n$ requires a particular type of AI system named *variational autoencoder* [6]. Usually, in complex dynamical systems, such autoencoders are used to extract the latent space and obtain the simplified differential equations defined on the latent space as entraining the full system. In other words, the autoencoder is used to simplify the dynamical system, retaining the dynamics that are essential and averaging out subordinated degrees of freedom. Chapter 3 will be devoted to describing such autoencoders, with a special focus on systems that yield topological metamodels.

In the context of extending the SM, or rather reverse engineering it to encompass a dormant dimension, we need to have the autoencoder working in reverse. Rather than simplifying the fields defined in latent space we need to lift them to a quintessential space of higher complexity, so as to recover the original fields when taking the quotient modulo the standard 4-dimensional space-time coordinates. In addition, we need the autoencoder to produce the

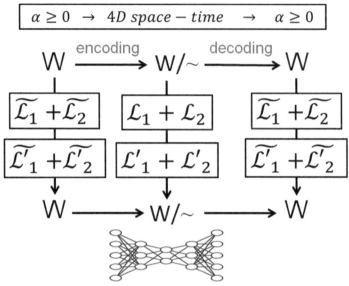

Variational autoencoder [compatibility⇄commutative diagram]

$$x^\mu \to x'^\nu \Rightarrow \int [\mathcal{L}_1 + \mathcal{L}_2]\, dx^\mu = \int [\mathcal{L}'_1 + \mathcal{L}'_2]\, dx'^\nu$$

FIGURE 2.12 Neural network with autoencoder architecture designed to reverse engineer the Standard Model by incorporating a fifth circular dimension.

topological metamodel that would enable the proper lifting, so that a diagram of the type presented in Figure 2.12 becomes commutative.

An illustration of the kind of tasks performed by the topological autoencoder of the SM is depicted in Figure 2.13. The figure describes the lifting to quintessential space W of a process, known as beta decay, comprising the neutron (n) transformation into a proton (p) by transformation of one of its constitutive down quarks (d) into an up quark (u), with emission of an electron (e^-) and an electron antineutrino $\left(\overline{v}_e\right)$. This process requires the interaction of the neutron with a W^--boson, a massive particle ($M = 80.433$ GeV/c^2) whose existence is ephemeral ($\tau \approx 10^{-26}$s~6.582×10^{-16}eVs/$(2 \times 80.433 \times 10^9$eV)), as its formation requires borrowing energy from the vacuum as a fluctuation governed by Heisenberg's uncertainty principle $\left(\Delta E \times \Delta t \sim \dfrac{\hbar}{2}\right)$. Here we adopt the notation $X = \left(n, W^-\right)$, $X' = \left(p, W^-, \overline{v}_e, e^-\right)$ to indicate respectively the initial and final state of the system as recorded on the latent space W/~ and dictated by the SM. The respective liftings to

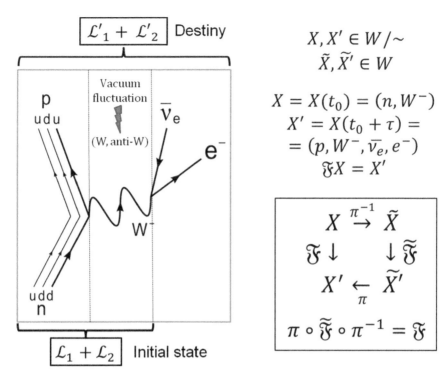

FIGURE 2.13 Elementary process of beta-decay defined on the latent space representing the 4-dimensional space-time W/~ and consistently lifted to the quintessential space W. The process consists of a neutron transformed into a proton through communication with a W⁻ gauge boson with emission of an electron and an electron antineutrino.

W are $\tilde{X}, \tilde{X'}$, and their validity is ensured if and only if they satisfy the flow relation:

$$\left(\pi \circ \tilde{\mathfrak{F}}\right)\tilde{X} = \mathfrak{F}X; \mathfrak{F}X = X'; \tilde{\mathfrak{F}}\tilde{X} = \tilde{X'}. \tag{2.14}$$

The operators $\mathfrak{F}$ and $\tilde{\mathfrak{F}}$ determine respectively all processes in W/~ dictated by the SM and the lifting of such processes to the quintessential space W. Thus, the operators may be assimilated to the time evolution propagators over time intervals identified with the lifetimes of the particles that communicate fundamental forces. This is a mere analogy but an autoencoder will not make the distinction in this context. Hence, the autoencoder is charged with the task of computing the operator $\tilde{\mathfrak{F}}$ and canonical projection π, so that the generic diagram in Figure 2.14 becomes commutative.

$$W \xrightarrow{\pi} W/\sim$$

$$\downarrow \widetilde{\widetilde{\mathfrak{F}}} \qquad \downarrow \widetilde{\mathfrak{F}}$$

$$W \xrightarrow{\pi} W/\sim$$

FIGURE 2.14 Commutative diagram for the quintessential lifting of a fundamental process in the reverse engineering of the Standard Model.

Similarly, the lifting of another fundamental process, the creation and decay of the Higgs boson (H) into two lepton-antilepton pairs is depicted in Figure 2.15. The creation of the Higgs boson entails the fusion of two gluons (g, g) mediated by a virtual top quark ($\hat{t}$), and the "timing" of the process is given by the sum of the lifetimes of the bosons (H, Z) that serve as intermediates.

In this way, the SM may be extended by the topological autoencoder to incorporate an extra dimension compatible with the compact and multiply connected topological metamodel of the universe validated experimentally

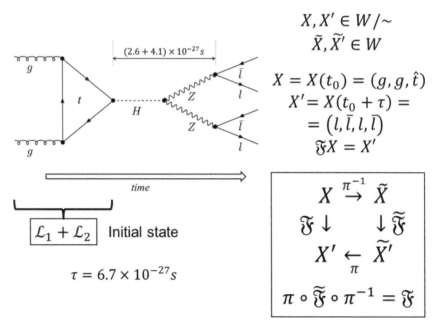

FIGURE 2.15 The lifting of the elementary process of creation and (detectable) decay of a Higgs boson.

by the spectral decomposition of the CMB. The topological autoencoder would then be in position to discover dark matter particles as it reverse engineers the SM to adapt it to the early universe where the dormant dimension became commensurate with the other four dimensions of space-time. This program will be carried out in Chapter 4.

2.7 AMALGAMATED HIGGS BOSON AS DARK-MATTER PORTAL

The SM is considered the towering achievement of particle physics, mainly because of its amazing predictive power and sparse parametrization. Despite its success, the SM in its current version does not include dark matter (DM). Furthermore, it seems unclear that it can be extended in a cogent way to encompass DM without significantly altering the existing models of observable particles and solving the quantum gravity and hierarchy problems. Ongoing research described in this book, particularly in Chapter 4, may turn this situation around. With the mathematical implementation of the concept of geometric dilution (v) within the premises of the SM, DM can be shown to percolate into observable dimensions. Therefore, we are confident that a suitable extension of the SM will be forthcoming. This discussion points to a viable possibility in this regard, as fermionic DM is shown to have a dynamical origin capable of interplay or mixing with the Higgs boson. What follows is an outline of the SM extension.

The approach requires a mixing of the scalar field for DM with the Higgs doublet scalar field to yield an irreducible mediator between the observable and dark sectors. This interplay has been modelled by imposing boundaries in the dark curly dimension, so the interplay may materialize at the 3-brane that represents the boundary. But such boundaries do not exist in the compact extra dimension, since it constitutes a circle. For this reason, we need to replace the topology of the dark dimension, $\mathcal{S}^1$, for the quotient manifold $\mathcal{S}^1/\mathbb{Z}_2$. This quotient is made up of orbits under the symmetry operation $\vartheta \to -\vartheta$, with the polar representation $y = r_0 e^{i\vartheta}$ of the dark coordinate. The metric in $\mathcal{S}^1/\mathbb{Z}_2$ is entirely determined by that of the subset $0 \leq \vartheta \leq \pi$ which yields boundaries, that is, the 3-branes $\vartheta = 0$, $\vartheta = \pi$. Furthermore, DM particles become $\mathbb{Z}_2$-odd scalar fields that can now be amalgamated with the Higgs scalar field to yield mediators between the observable and dark sectors. The lightest such portal corresponds to the combination with the first Kaluza-Klein (i.e., at $v = 0$) excitation of a dark fermion, which corresponds to the ur-Higgs.

This construction of the compound particle departs from the parametrized semiempirical model of the Higgs boson, with its quartic term in the potential energy accounting for self-interactions of the Higgs field $\phi_h = \phi_h(x, t)$. In the SM, the spontaneous symmetry breaking in the Higgs potential, responsible for endowing particles with mass, is associated with the nonzero vacuum expectation value (v.e.v.) v arising from self-interactions that yield the potential term $-\lambda\phi_h^4$ with coupling constant λ. The parametrization of the v.e.v. at $v = 246$ GeV is adjusted to yield all particle masses. By contrast, our model does not include self-interactions; instead, it extends the field ϕ_h to the amalgamated field $\phi_H = \phi_H(x, y, t)$ by incorporating a $\mathbb{Z}_2$-odd DM-field $\phi_D(y)$, that spans the dark dimension:

$$\phi_H = \phi_H\left(x, y, t\right) = \phi_h\left(x, t\right)\left[1 - \left(\frac{\phi_h\left(x, t\right)}{\sqrt{2}\,\phi_D\left(y\right)}\right)^2\right]^{1/2}. \tag{2.15}$$

The Higgs field thus aquires the v.e.v. stored in the dark dimension with quotient topology $\mathcal{S}^1/\mathbb{Z}_2$, while only the (x, t)-dependence of the Higgs field is apparent in the SM (in its current version).

By adopting the radial representation $\phi_h = \rho e^{i\omega}$, the Higgs boson becomes, in standard space-time, an oscillatory excitation of the radial field $\eta = \rho - v$, while the Lagrangian for η lifted to the (x, t, y)-5-dimensional space-time becomes

$$\widetilde{\mathcal{L}\left(\eta\right)} = \frac{1}{2}\left(\partial\eta\right)^2 - \lambda\overline{\phi_D}\phi_D\eta^2. \tag{2.16}$$

This result implies that the Higgs boson mass $m_H = \sqrt{2\lambda}v\sqrt{2\lambda} = \cos\alpha$, assimilated to geometric dilution parameter is bestowed through a coupling with the DM scalar field. If geometrically undiluted, the v.e.v. thus stored would constitute the lightest DM particle, namely the ur-Higgs. Equations (2.15) and (2.16) inform that DM can be incorporated into the SM via the key observation that the v.e.v. of the Higgs field is actually a $\mathbb{Z}_2$-odd stationary matter-wave spanning the dark circular dimension. This circle has radius $r_0 = 0.802 \times 10^{-18}$ m, which corresponds to the Compton length of the quark [4], the smallest material dimension.

In this model, the amalgamated Higgs field (Equation 2.15) enables the spillover of the dark universe into the observable universe, acting as gatekeeper and portal between the observable and dark sectors. This is because the parameter that embodies the geometric dilution $v = -\log \cos \alpha$ of the ur-Higgs onto observable dimensions to yield a specific particle mass (Chapter 4) can be unambiguously associated with the Yukawa coupling, g_Y, of the fermion field with the Higgs field, and with the mass-to-v.e.v. ratio in the case of the weak-force bosons. *The behavior of the amalgamated Higgs boson as portal for DM, enabling the spillover that endows observable fermions with mass, is enshrined in the fundamental proportionality*

$$g_Y \propto \cos \alpha. \tag{2.17}$$

Under these observations, it becomes clear that the amalgamated Higgs boson serves as the portal of the dark universe, with a controlled spillover mechanism that endows SM particles with mass.

The topology $S^1/\mathbb{Z}_2$ assigned to the dark dimension introduces an artificiality that gets justified a posteriori. It enables physicists to address the hierarchical problem whereby gravity is extremely weak compared with the three other forces and lies outside the brane that enshrines the SM. Unlike S^1, the topology of $S^1/\mathbb{Z}_2$ presents boundaries/branes, while enabling storage of momentum as stationary matter waves. The relativistic metric is concentrated at $\vartheta = 0$, the gravity brane, while the SM is enshrined in the $\vartheta = \pi$ brane. Extending the SM to incorporate the dark sector requires creating an autoencoder that distills the physics from $W_4 \times S^1$ into $W_4 \times S^1/\mathbb{Z}_2$. Thus, the arbitrariness of the dark quotient space is removed while its modeling virtues are retained.

REFERENCES

1. Weinberg S (2008) *Cosmology*. Oxford University Press, New York, USA.
2. Feynman RP, Weinberg S (1999) *Elementary Particles and the Laws of Physics*. Cambridge University Press, Cambridge, UK.
3. Profumo S (2017) *Introduction to Particle Dark Matter*. World Scientific Publishing, Singapore.
4. Abramowiczy H, Abtt I, Adamczykh L, Adamusae M, Antonelli S, et al. Zeus Collaboration (2016) Limits on the effective quark radius from inclusive *ep* scattering at HERA. *Phys Lett B* 757: 468–472.
5. Hawking SW, Ellis GFR (2023) *The Large Scale Structure of Space-Time: 50th Anniversary Edition*. Cambridge University Press, Cambridge, UK.

6. Fernández A (2022) *Topological Dynamics for Metamodel Discovery with Artificial Intelligence.* Chapman & Hall/CRC, Taylor and Francis, London, UK.
7. Aurich R, Buchert T, France MJ, Steiner F (2021) The Variance of the CMB Temperature Gradient: A New Signature of a Multiply Connected Universe. *Class Quant Grav* 38: 225005.

CHAPTER 3

Methods

Quintessential Autoencoders

"The physical world is only made of information,
energy and matter are incidentals."
JOHN A. WHEELER

This chapter introduces AI systems at a fairly elementary level and broadly delineates the possibilities of AI for model discovery, focusing on dynamical systems. The presentation deals mostly with deep learning, autoencoders, and other more specialized architectures and is tailored to researchers seeking to unravel physical models distilled from big data. With the leveraging of artificial intelligence, dynamical systems have found a fertile ground for development. Machine learning identifies parsimonious models providing physical underpinnings of time-series data. However, such heavily parametrized models hardly yield physical laws. The problem becomes daunting as we turn to multi-scale complexities. This chapter addresses these imperatives as it introduces topological methods that enable metamodel discovery and the proper computational tools to decode the metamodel as an inferential framework. The methods advance model discovery, enabling reverse engineering of big data arising in a wide range of cosmological applications, where metamodels with emergent quantum behavior are crucial to provide physical underpinnings of quantum gravity.

The approach is mainly directed at identifying parsimonious metamodels of dynamical systems that describe complex contexts.

DOI: 10.1201/9781003501534-4

This chapter also addresses the problem of quantum gravity as an emergent property in the physics of machine learning. To that effect, the chapter explores the possibility of an AI–based construction of a quantum holographic autoencoder, which requires that we first deal with the physics of machine learning and specifically enquire whether emergent quantum behavior can arise in a neural network. By emergent quantum mechanics we mean a formulation within a framework of nonlocal equilibrated hidden variables, as in the Bohm scheme. Once an emergent quantum behavior is shown to become possible within the machine learning system equilibrated on the nontrainable – that is, hidden – variables, we address the question of developing a relativistic string gravitational scheme on the hidden variables adopted. Thus, the network with equilibrated nontrainable variables becomes in effect a quantum gravity autoencoder for the network exhibiting emergent gravity in the nonequilibrium regime prior to the equilibration of the nontrainable variables. In this way we build a quantum metamodel for gravity that fulfills at least in part a major imperative for physicists seeking a unified field theory.

Finally, by coupling two physical embodiments of quantum-gravity autoencoders, we implement a cosmic technology for universe reproduction through quantum tunneling across the interface between the two autoencoders. The technology is supplemented with a space-time rendering of the quantum entanglement of the two primeval wormholes associated with the creation of a progeny universe through quantum tunneling.

3.1 PRIMER ON NEURAL NETWORKS FOR MODEL AND METAMODEL DISCOVERY

The leveraging of artificial intelligence (AI) for model discovery in dynamical systems is revolutionizing both disciplines, leading to a mutually beneficial development. The cases of interest in this book are not amenable to model discovery in the sense of yielding a sparse system of differential equations that entrains the full dynamical system. Rather, we are seeking for something more elementary and subtle: A metamodel, or topological representation of the dynamics. With the implementation of topological methods, AI–empowered metamodel discovery is able to focus on levels of system complexity and multi-scale hierarchies considered off limits for current AI technologies. The information on time series is encoded at the maximum level of coarse graining, hence greatly simplifying the computations while enabling a decoding of the information generated at the level of a topological description.

In dealing with dynamical systems using AI–based approaches, we address the following core question: What constitutes an insightful parsimonious model? The standard answer is: "A sparse system of differential equations on latent coordinates." As argued in this chapter, this is not necessarily the format chosen by AI, given the "dimensionality curse" associated with the ultracomplex realities we chose to work on. The deployment of AI requires a paradigm shift, where dynamic information gets encoded in what would be termed a "topological metamodel." The metamodel is essentially pattern-based, where AI–interpretable topological patterns encode physics laws. These methods are likely to advance model discovery as they enable the reverse engineering of time series stemming from vastly complex realities hitherto inaccessible to other AI methods.

AI refers to machines capable of exhibiting behavioral traits that humans regard as indicators of intelligence, such as learning and problem solving [1]. Within this protean subject, machine learning (ML) refers to the ability to learn without being explicitly instructed to do so, while deep learning (DL) refers to an automated extraction of features, patterns, and ultimately models from arrayed data that is sequentially represented within an abstraction hierarchy organized as a multilayered neural network (NN) [2–4].

DL has been shown to be highly efficacious at identifying features that are in principle discoverable from the data [2, 5]. As in face recognition, features are hierarchically organized, so that large-scale patterns (eyes, noses, face shapes) emerge after several layers of abstraction from simpler or more rudimentary patterns (lines, curves, shades). The beauty and power of DL resides in the fact that the feature extraction process may be carried out in an unsupervised manner: The features emerge from the training of the system without human input or bias and enable the network to make accurate inferences. In this era of big data, we may state that there are several compelling reasons for implementing DL approaches:

- Fields like biology, particle physics, and cosmology are generating vast amounts of data and time series that can be easily stored and interpreted to achieve a conceptual unification within overarching models,

- we have the right hardware, that is, graphics processing units (GPUs) that are massively parallelizable, and

- we have adequate software such as TensorFlow (TF) that enables suitable modular coding if the data can be pixelized or voxelized into a tensorial array, be it a vector, a matrix or a tensor proper [3, 4].

At the most basic level, the building block of an NN is the neuron, referred to as *perceptron* [5]. The perceptron enables forward propagation of information encoded in an array of inputs x_1, x_2, ..., x_n weighted by parameters w_1, w_2, ..., w_n to generate an output of the form $y = f\left(\sum_{i=1}^{n} x_i w_i + w_0\right)$ where w_0 may be regarded as a bias term and f is a nonlinear activation function, often a sigmoid or sigmoid shaped, as shown in Figure 3.1. The bias term enables to shift the activation function. In vector representation, we often write: $y = f(z)$, $z = x.w + w_0$, where x,w are respectively input and weight vector. One can subsequently build a fully connected layer of perceptrons indexed by j, whereby $z_j = x.wj + w_{0,j}$. The dense layer can be readily implemented in TF code by simply specifying number of outputs/perceptrons [3, 4]. We usually refer to the vector of linearly transformed inputs z as "hidden layer," as it does not explicitly describe observables [5].

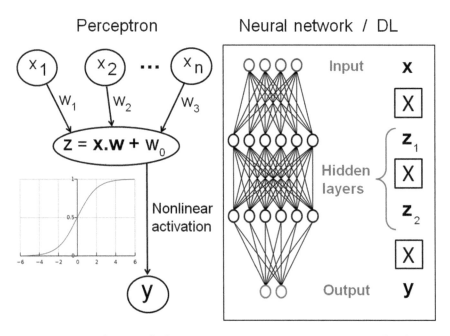

FIGURE 3.1 Scheme of the perceptron or neuron activation by linear transformation of input stimulus (*x*) followed by nonlinear signal transmission as output *y*. The panels on the right show the organization of NNs with one and multiple hidden layers. The "boxed X" indicates full (dense) node connectivity between consecutive layers.

Hidden layers may be stacked as in DL architectures. Thus, for the j-th perceptron in the k-th hidden layer, we get: $z_j^{(k)} = \sum_{i=1}^{n_{k-1}} y_i^{(k-1)} w_{ij}^{(k)} + w_{oj}^{(k)} = \sum_{i=1}^{n_{k-1}} f\left(z_i^{(k-1)}\right) w_{ij}^{(k)} + w_{oj}^{(k)}$. The sequential composition of the network by stacking hidden layers has a standardized script in TF that generates the propagation of information as specified by the preceding equation[3, 4]. In an NN with K hidden layers, we may regard the output $y_j^{(K)} = f\left(\sum_{i=1}^{n_{k-1}} y_i^{(K-1)} w_{ij}^{(K)} + w_{oj}^{(K)}\right)$ as an "inference" made by the DL system.

The accuracy of the DL inference represents the level of optimization of network performance and may be assessed vis-à-vis a training set of input/output paired data points. This assessment is often referred to as the *loss* of the network [2, 5]. The loss is parametrically dependent on the full weight tensor $W = [w_{i_{k-1} j_k}^{(k)}]_{k=1, ..., K}$ which may contain a huge number of weights, in the thousands if not millions, depending on the size of the network. Thus, for DL NN with K hidden layers the loss function or empirical risk $J(W)$ becomes $J(W) = |\mathfrak{I}|^{-1} \sum_{\xi \in \mathfrak{I}} \mathcal{L}\left(y_\xi^{(K)}(W), y_\xi\right)$, where $\mathfrak{I}$ is the training set with number of elements $|\mathfrak{I}|$, $y_\xi^{(K)}(W)$ is the predicted output vector for input vector $x_\xi, \xi \in \mathfrak{I}$, y_ξ is the actual output vector, and $\mathcal{L}\left(y_\xi^{(K)}(W), y_\xi\right)$ measures the discrepancy between actual and predicted output. In regression problems, where the output is a numerical vector, it is often convenient to adopt $\mathcal{L}\left(y_\xi^{(K)}, y_\xi\right) = \left\|y_\xi^{(K)}(W) - y_\xi\right\|^2$. In such cases the optimization of the NN through training becomes a problem of least squares. To optimize the NN is tantamount to minimize the loss $J(W)$, which requires a careful fine-tuning of the size of the training set vis-à-vis the size of the weight parametrization. In principle, the optimal network is defined as: $W = W^* = \text{Argmin } J(W)$.

An insufficient number of training input/output pairs relative to the size of the weight tensor would give rise to *overfitting*, requiring special techniques, generically known as *regularization*, in order to trim the network, that is, randomly remove connections, without compromising predicting efficacy [2, 5].

Optimizing the network involves the laborious and costly computation of the minus gradient $-\dfrac{\partial J(W)}{\partial W}$, which locally indicates the direction of steepest descent in the multidimensional surface $J = J(W)$. An iterative gradient descent computation generating a fine-tuned weight updating $W \to W - h \dfrac{\partial J(W)}{\partial W}$ should eventually lead to convergence to a local minimum of $J = J(W)$ when adopting a suitable learning step h. This parameter should be tuned to effectively escape local minima while avoiding over-shooting in trying to reach the global minimum. Most gradient descent algorithms use an adaptive learning step during training, in accord with the constraints indicated. In practice the gradient problem is approached by what is called the *stochastic gradient descent* method, whereby not all datapoints (input/output pairs) in the training set are used in each mini-mization step but the gradient is approximated by an average over ran-domly chosen batches of datapoints in a tradeoff between accuracy and computational efficiency. Obviously, batch size and learning rate are cor-related, so the more accurate the gradient estimation, the larger the learn-ing step may be (a token of computational confidence). To achieve significant speed, the stochastic gradient descent computation may be massively parallelized by splitting up batches into multiple GPUs.

3.2 NEURAL NETWORKS AS DYNAMICAL SYSTEMS

In this section we shall be concerned with data organized as a time series arrayed as $\{x(t_0), x(t_0 + \tau), x(t_0 + 2\tau), ..., x(t_0 + L\tau)\}$, where $x(t)$ is the vec-tor of observables at time t and τ is the interval that determines the time coarse graining inherent to the sequential detection registered in the vec-tor x. The time series enables a training of the NN such that the output vector $y = y(x(t), W)$ should approximate $x(t + \tau)$ when the input is $x(t)$, and this correspondence is carried over all t in the training time series. Thus, for a given network architecture, the optimal network is the one that realizes the minimum of the loss function:

$$J(W) = \sum_{n=1}^{L} \left\| x(t_0 + n\tau) - y(x(t_0 + (n-1)\tau), f, W) \right\|^2. \qquad (3.1)$$

Obviously, for a fixed activation function, a model for the time series may be given simply by Argmin $J(W)$, but such a model would lack universal-ity as it would be extremely parametrized, most likely overparametrized, and would not prove insightful in the sense that it is not parsimonious. The discussion prompts us to enquire what truly constitutes a model. This question will be addressed subsequently.

The most common time series that humanity has collected since time immemorial stems from astronomical observation. For about two millen-nia until the time of Copernicus, and Newton later on, humanity had been striving to find a suitable model that would fit and explain the data, that is, the recorded sequential positions of a set of celestial bodies. In today's more general context, biology, particle physics, and cosmology are generating dynamical data at a staggering rate, while models that fit and explain the data are sorely lacking or hopelessly inconsequential. It is expected that the advent of AI will dramatically impact this sort of model discovery.

In essence we seek for what is known as autoencoder, an intermediate output with a dimensionality reduction and a simplified discerning physical picture that should therefore prove insightful to make sense of the patterns enshrined in the dynamics, enabling meaningful output inferences. These autoencoders and the models they give rise to will be studied in detail in the subsequent chapter.

3.3 DEEP LEARNING AND CONVOLUTIONAL NEURAL NETWORKS FOR FEATURE EXTRACTION

The huge output of biomedical and biostructural data have become the hallmark of the post-genomic era while chemical combinatorial possibilities make it forbiddingly difficult to parse chemical space in search for suitable leads for targeted therapy [6]. In this scenario, pharmaceutical researchers have turned to DL for guidance in drug discovery and development and target validation [7–10]. Thus, pharmacoinformatics has benefited immensely from the advent of DL systems trained to pair chemical compounds with molecular attributes likely to have therapeutic impact. These computational and informational techniques enable the evaluation of chemical compounds for specific properties, including target affinity, affinity screening profiles, structural features of drug-target docking, and ADMET (absorption, distribution, metabolism, excretion, and toxicity) profiling [8].

A major challenge in implementing DL models for pharmacoinformatics in accord with the generic scheme outlined stems from the need to represent the chemical structure as a tensor array (vector, matrix, or tensor proper) of pixelated or voxelated inputs that may be subsequently interrogated geometrically across the hidden layers of NN in search for features that are indicative of the molecular properties indicated. There are a number of representations of chemical space amenable to TF encoding. The most obvious one is to order the atoms in the compound on a 1D-array (following, for example the IUPAC numerical labeling convention) and represent the chemical structure of the molecule as a covalent bond matrix pairing atoms in row and column in accord with their covalent linkages, including single, multiple, and resonant (aromatic) bonds.

The matrix is subsequently transformed into a topological descriptor that describes the invariants arising from different atom ordering. The input describing chemical structure is paired within a training set against the molecular attributes of therapeutic relevance that the network is meant to infer. Then, compounds that need to be evaluated/profiled are inputted as the array of pixels/voxels, transformed into a topological representation, and feature extraction is achieved through the sequential activation of hidden layers at increasing levels of abstraction, eventually leading to the profile inference that is subsumed in the output layer.

Feature extraction through the NN often requires particular architectures known as convolutional NNs (CNNs) [2–5]. The idea is to pixelize or voxelize the input data in a matrix or 3D tensor array and then scan (convolve) the array with a filter associated with a specific pattern to generate a feature map. For the sake of the argument, let us consider a 2D-array input M. The filter $F=(w_{ij})$ is an mxm matrix actually representing a convolutional kernel, so that a neuron in the F-associated hidden layer M_F only senses the pixels in an mxm patch (receptive field), and the layer becomes the feature map

$$M_F = M * F = \left(\sum_{j=1}^{m} \sum_{i=1}^{m} w_{ij} x_{i+ap\ j+bp} + w_0 \right)_{ab} , \qquad (3.2)$$

where p, usually set at $p = 2$, is the stride adopted as the filter slides along the input matrix array, with dummy integers a, b indicating the patch location. Essentially, the CNN is an NN where the set of weights in each mxm patch of inputs are always the same as the filter slides along the input array to generate the hidden layer that constitutes the feature map. The convolution operation becomes the entry-by-entry (Frobenius) matrix inner product as the filter slides along the input matrix with a given stride. Successive filters may be applied, reducing progressively the size of the feature maps as higher and higher levels of abstraction are achieved in the representation of the data (Figure 3.2). Thus, the parametrization required for feature extraction in CNNs is relatively small, as all the neurons in a hidden layer share the same connectivity parameters that define the filter that generated that layer.

In CNNs the filters are not specified *a priori*, that is, their weights are not fixed through human intervention. The filter parametrization is automatically determined by the training/optimization process without

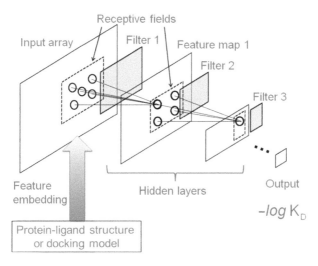

FIGURE 3.2   Scheme of a CNN used for inference of drug-target affinity through sequential application of filters that become optimized through minimization of the network loss.

introducing any assumption, other than the size of the receptive fields for each convolution operation and the overall number of filters to be applied. Thus, feature extraction in a CNN is carried out in an unsupervised manner and only requires that we script (in TF coding) the number of hidden layers or feature maps and the size of the filters to be applied to generate each hidden layer. The features themselves emerge as the network is trained.

Thus, CNNs often constitute AI–empowered platforms for drug discovery, where the structure of a protein-ligand interface that serves as precursor for a predicted drug-target interaction is pixelated in a feature embedding process as a 3D spatial array of protein-drug atom pairs deemed to be interacting across the interface [11]. The inference of target affinity for a given drug is assessed through a sequence of feature extractions using convolution filters until the output feature map becomes a number directly associated with the affinity $pK_D = -\log K_D$, where K_D is the dissociation constant for the drug-target complex inputted via feature embedding (Figure 3.2). The training of the network is carried out by minimizing the network loss or empirical risk over a set of drug-target complexes whose structure and affinity are both known (preferably, experimentally determined). The proteins in the complexes of the training set are typically homologs of the one whose affinity for a specific drug we seek to infer, so that the features that enable the affinity inference may emerge from structural alignment.

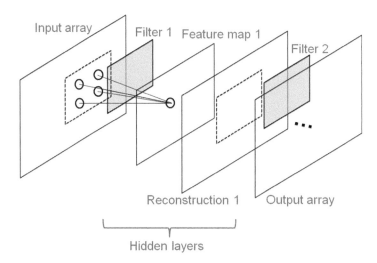

FIGURE 3.3   Scheme of a CNN with a particular type of architecture in which the patches that yield the pixels with the highest weights in the feature map are recreated as filter-stenciled "perfect" features. Thus, a feature-enriched reconstitution of each layer is carried out prior to the application of the next filter.

In other, more complex applications of CNNs, where the output is not simply a numerical parameter but rather a numerical array, it is often convenient to adopt a variation of the CNN architecture in which the patches that yield the pixels with the highest weights in the feature map are reconstituted into features perfectly stenciled by the filter, while other patches that do not yield discernible features are left invariant in non-overlapping regions. Thus, with the application of each filter, a feature-enriched reconstitution of the input layer is carried out prior to the application of the next filter (Figure 3.3).

3.4 AUTOENCODERS SIMPLIFY THE DYNAMICS IN LATENT MANIFOLDS AND QUOTIENT SPACES

It is widely expected that AI, and DL in particular, will become a major player in model discovery for data-driven research [12]. Within this vast array of possibilities, the focus of this book is model discovery for dynamical systems that underlie big time-series data from vast areas as distant as biomedicine and cosmology [13–15]. The biggest hurdle we stumble upon is that the level of complexity of the data generated in fields like biology and cosmology is so extreme that it challenges the notion of model itself. As shown subsequently, the model itself can seldom be cast as a system of differential equations. Thus, much of the ensuing discussion deals with the question of what constitutes a meaningful model with predictive value when dealing with the ultracomplex realities represented in the contexts of biology or cosmology.

Encoding has proven to be a necessary category in AI [12–15]. The type of AI we are mostly concerned with involves deep learning, which requires an encoding of the raw information that needs to be acquired and processed further to make meaningful inferences. Just like with human intelligence, the encoding problem stems from the core question: What is essential and what it superfluous? The encoding problem becomes solvable when the system under scrutiny is hierarchical and the hierarchical structure is fairly obvious or at least discoverable. In the cases treated in this book, the structure of the data is always hierarchical, implying that the learning process admits a reductive approach represented by the encoding. When the network architecture is such that this process is automatically generated, we name the NN *autoencoder*.

We shall deal mostly with time-dependent data representing physical or biophysical systems, where detailed fast motions may be systematically averaged out, so the relevant information may be stored in a coarse-grained representation. Rigorous mathematical constructs will be introduced to implement the hierarchical encoding materialized by the autoencoder. Some examples of hierarchical encoding of physical or biophysical processes that needs to be taken into account when designing the autoencoder architecture are:

1) The adiabatic approximation, where fast-relaxing or fast-evolving enslaved degrees of freedom are averaged out, or thermalized or equilibrated when incorporated into a model for the time evolution of a dynamical system [16]. In molecular physics, examples of such degrees of freedom are vibrational hard modes that evolve on timescales of the order of picoseconds to nanoseconds, while soft modes evolve on longer timescales ranging from the submicrosend to seconds.

2) In atomic physics, the Born-Oppenheimer approximation represents regions of the potential energy surface where an adiabatic regime holds so the motion of electrons is enslaved or entrained by the slower motion of the atomic nuclei [17].

3) The latent manifold in dynamical systems [12], where the system is entrained or enslaved in the long-time limit by the evolution in

a manifold of lower dimension than the original space. The latent manifold is often referred to as center manifold [16], especially in the context of dissipative systems, and contains the attractors of the system (Figure 3.4).

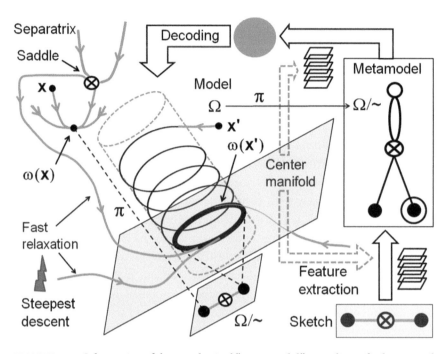

FIGURE 3.4 Schematics of the topological "metamodel" encoding of a dynamical system that contains a center manifold. The center manifold enslaves or entrains the dynamics for timescales associated with the adiabatic elimination of fast-relaxing and thermalized degrees of freedom and therefore constitutes a latent manifold (Ω) within which a model (differential equations on latent coordinates) may be identified by an autoencoder. The dynamic information may be encoded further by a second autoencoder at a higher level of abstraction, where the dynamics are represented more coarsely as "modulo basin" transitions. The modulo-basin dynamics is mapped on a quotient space, $\Omega/\sim$, where two states x, y are regarded as equivalent, $x \sim y$, if they have the same destiny state ($\omega(x) = \omega(y)$). The encoding processes are symbolized by dashed lines, and the enslaving center-manifold dynamics is highlighted by a thick dark circle. Thus, the second autoencoder materializes the projection $\pi: \Omega \to \Omega/\sim$ and represents the dynamics as a walk in a graph whose vertices are the critical points (minima, saddles of different indices, maxima) and attractors of the system, and the edges represent connections along pathways of steepest descent. This topological metamodel may be subsequently decoded back to flesh out the dynamical system using learning technology.

4) In the biophysical context of functionalized soluble proteins (enzymes), the quantum mechanics for specific chemical processes take place at the *epistructure* (solvent organization around the protein structure), where water is chemically functionalized [18]. A rigorous treatment of this problem would require that we solve a time-dependent Schrödinger equation. In this context, an AI approach would need to be incorporated to learn to infer the nodal structure of molecular orbitals within a voxelated 3D grid. With current technologies, this AI approach is plausible only under the adiabatic regime given by the Born-Oppenheimer approximation.

5) In clonal population dynamics, cancer phenotypes that become selected under therapeutic pressure [19]. This dynamics is complex but hierarchical, and the mathematical procedure to "encode what is essential" in this context is based on the center manifold reduction (Figure 3.4). This reduction discussed in 3) is actually a projection of the dynamical system onto a lower dimensional system that entrains it. The autoencoder may discover the center manifold by interrogating vast amounts of time-dependent data, as it seeks to meaningfully reduce dimensionality in such hierarchical systems.

6) Chapter 4 introduces the *quotient space*, a fundamental mathematical construct to take advantage of dynamic hierarchy in order to encode information as required to implement DL systems [20]. The quotient space is built upon an underlying dynamics and may be equated with the orbit space, that is, points in the same trajectory are regarded as equivalent, and the quotient space is the set of equivalence classes with a topology inherited from the underlying space where the dynamical system is mapped (Figure 3.4). In rigorous terms, two points-states (x, x') are equivalent $(x \sim x')$ or belong to the same equivalence class $\left(x' \in \bar{x}, \overline{x'} = \bar{x} \right)$ if and only if they share the same destiny state $(\omega(x) = \omega(x'))$ vis-à-vis the trajectory to which they both belong. Thus, we simplify the space by lumping microstates into basins of attraction (of destiny states) in the potential energy surface. The modulo-basin dynamics constitutes a coarse model named *metamodel*, which is far easier to encode as the system learns data generated by atomistic molecular dynamics simulations. In this way, AI learns to propagate dynamics in quotient space, discovering a metamodel to cover physically realistic timescales usually inaccessible to detailed atomistic computations.

While the center manifold encoding is suited for dissipative dynamics, where the attractor may be nontrivial (cf. Figure 3.4), the quotient space simplification is better suited for Hamiltonian systems. Both levels of encoding converge as free-energy dissipation tends to zero. In fact, as the AI system encodes time-dependent raw data, it implicitly composes the two levels of encoding, with the center-manifold reduction averaging out of fast modes, followed by projection onto quotient space (Figure 3.4).

In all cases dealt with in this book, the hierarchy of the data that makes it amenable to encoding is either AI–discoverable, or it may be unraveled through a rigorous mathematical construct that needs to be incorporated to the learning code, as shown in the following.

The modulo-basin hierarchical representation of the dynamics in quotient space enables the construction of metamodels, that is, coarse-grained models that represent transitions between equivalence classes of states of the system, essentially reproducing the topological dynamics within a graph representation (Figure 3.5). As an illustration, let us consider the "keto" $\Longleftrightarrow$ "enol" interconversion of acetone (Figure 3.5). This is a chemical reaction involving an intramolecular proton migration that can be modeled by the potential energy surface (PES) representing the ground-level electronic energy sheet under the Born-Oppenheimer approximation. From quantum mechanics calculation we know that this PES has three minima, representing the less stable "enol" form (1), the more stable "keto" form (2), and a state where the jumping proton is completely dissociated from the rest of the molecule (3). The minima are separated by four saddle points at the top of the path of steepest descent joining pairs of minima on both sides of the saddle along the direction of negative curvature. This direction is actually the eigendirection associated with the eigenvalue of the Hessian matrix (Jacobian of the gradient vector field) with negative real part computed at the saddle point. In turn, the lines of steepest descent joining the four saddles with the three maxima become the separatrices of the basins of attraction of the minima. Thus all lines of steepest descent may be represented as a graph (metamodel II, Figure 3.5) joining critical points that are in turn organized in tiers, where the lowest tier corresponds to points where all eigendirections have positive curvature (minima) and the next tier includes all critical points with only one eigendirection with negative curvature (saddles of index 1), the next tier is associated with two eigendirections of negative curvature (maxima in the case of Figure 3.5), and so on. The edges on the graph represent paths of steepest descent joining critical points in adjacent tiers. In this way the

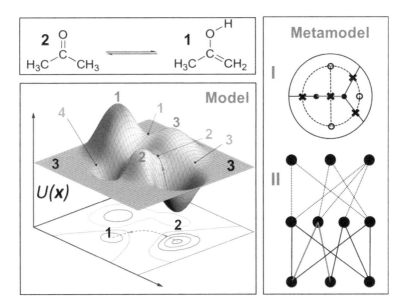

FIGURE 3.5 Three levels of abstraction in the modeling of the chemical dynamics for the isomerization "keto" ⟺ "enol" of acetone, representing the intramolecular migration of a proton. The potential energy surface represents the model and the topological relations between critical points, together with the "modulo basin" transitions along the paths of steepest descent joining critical points, represent the metamodel. On metamodel I, the modulo-basin topology is mapped onto the latent manifold, while on metamodel II the modulo-basin topology is abstracted further and displayed as a graph.

topological dynamics of the PES may be encoded as a metamodel represented in graph form, while the reversible chemical reaction pathway becomes a walk in the graph.

As argued subsequently, metamodels enable the discovery of hierarchical dynamical systems underlying processes that unravel in realities of high multi-scale complexity. The implementation of a metamodel factorization of the dynamics requires the concerted participation of several components operating in a coordinated manner within an AI platform. First, we need to introduce the so-called autoencoders that constitute the deep NN systems that encode the dynamics on the "latent manifold" Ω. This manifold entails a significant dimensionality reduction relative to state space W and is spanned by the internal coordinates that label the orbits of the symmetry group $\mathcal{G}$ inherent to the system and acting on W: $\Omega \approx W/\mathcal{G}$. Traditionally, it is expected that the first autoencoder, hereby denoted AE1, that generates the latent manifold is jointly optimized to generate also the most sparse or minimal set of differential equations on the manifold that can be decoded back onto the dynamical system defined on W. This is what is usually meant by "model discovery." In practice, the level of multi-scale complexity of the processes dealt with in this book does not make models amenable to discovery at the geometric level. As previously argued, in such cases another level of abstraction $\pi : \Omega \rightarrow \Omega/\sim$ needs to be introduced so that the coarse graining of time within $\Omega/\sim$ reflects equilibration within the basins of attraction in the latent dynamics. This hierarchical escalation in the level of abstraction requires a second autoencoder, AE2, capable of encoding the topological features of the latent vector field.

At this stage a different sort of NN architecture is required to propagate the metadynamics on $\Omega/\sim$. To properly delineate the architecture of the NN required for metadynamic propagation, we limit the discussion to the case where the dynamics is generated by a smooth (i.e., C^1) potential energy function $U : W \rightarrow \mathbb{R}$ invariant upon the isometries (distance-preserving transformations) of W. Furthermore, $\mathcal{G}$ is a Euclidean group, so that Ω is compact and hence $\Omega/\sim$ becomes a discrete set of basins, and a basin assignment represents a coarse state of the system. Then, the metadynamics may be encoded as "evolving text" representing a time series of basin transitions. The textual processing requires the implementation of a particular type of DL architecture known as *transformer*, while the transformer-based propagation of the metadyamics constitutes a Markovian process (Figure 3.6). Through AE2, this metadynamics is decoded as latent dynamics on Ω and validated by contrasting the latent dynamics against the hidden Markov process upheld under the adiabatic conditions described.

Thus, to implement a metamodel within an AI platform, it is essential that the two autoencoders and the transformer are optimized to work concertedly with complete compatibility (Figure 3.6). This means that an input state yields the same destiny state regardless of the pathway chosen provided the pathways have identical endpoint spaces in the commutative diagram presented in Figure 3.6.

3.5 REVERSE ENGINEERING OF THE STANDARD MODEL WITH AN AUTODECODER

The notion of diagram commutativity (Figure 2.12) is central to the encoding of a dynamical system into its latent dynamics and, reciprocally, to the decoding of the latent dynamics onto the full dynamics. Thus, the task of

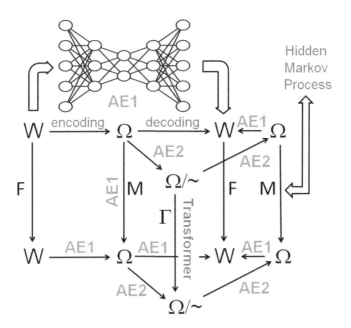

FIGURE 3.6 Commutative diagram representing two coupled autoencoders operating sequentially in tandem and representing two levels of abstraction of a dynamical system. The first autoencoder, labeled AE1, projects the dynamics onto the latent manifold, while autoencoder AE2 projects the latent dynamics onto a discretized "modulo basin" version where a coarse state of the system is represented in textual form and the "modulo basin" dynamics may be learned and propagated within a special type of autoencoding architecture known as "transformer."

decoding the Standard Model, canonically defined on the 4-dimensional space-time $W/\sim$, onto a 5-dimensional compact multiply connected manifold W requires an autoencoder operating in reverse. This reverse autoencoder lifts the flow $\mathcal{F} : W/\sim \to W/\sim$ to a flow $\tilde{\mathcal{F}} : W \to W$ that must be compatible in accordance with the commutativity condition: $\mathcal{F} \circ \pi = \pi \circ \tilde{\mathcal{F}}$, with $\pi : W \to W/\sim$ denoting the canonical projection that assigns each point in W to its equivalence class in $W/\sim$. The key issue with this formulation is that the Standard Model *per se* does not represent a dynamical system in any obvious way, so what sort of flow are we actually discussing in this context? This question is crucial because we are introducing AI technology that is tailored for dynamical systems.

Fortunately, we may treat the interaction processes described by the Standard Model as transformations created by a time-dependent propagator. For each elementary process, this operator has a time step associated with it, and this time step is exactly the lifetime of the boson that communicates the force in the process of particle transformation, as obtained from Heisenberg's uncertainty principle: $\sim \dfrac{\hbar}{2M_{\text{B}}c^2}$, where M_{B} is the rest mass of the respective gauge boson. There may be more than one boson involved, as in the case of creation and decay of the Higgs boson through gluon fusion, in which case the time step is the sum of the respective lifetimes. The dedicated AI technology of variational autoencoders may be thus leveraged to reverse engineer the standard model, a process tantamount to identify the lift flow $\tilde{\mathcal{F}} : W \rightarrow W$ for a propagator $\mathcal{F} : W/\sim \rightarrow W/\sim$ generically determining an elementary process in dynamic form. In this setting, "reverse engineering" is equated with identifying the lift propagator that would make the diagram in Figure 3.7 commutative.

The AI technology described in this chapter will be applied in reverse in Chapters 4 and 5 in order to identify dark matter and dark energy in 5-dimensional space-time. This is tantamount to reverse engineer the Standard Model by incorporating the dormant dimension shown to store gravity.

Reverse engineering of the Standard Model

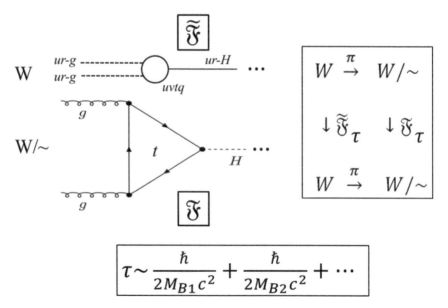

FIGURE 3.7 Diagram commutativity fulfilled by the autoencoder that carries out the reverse engineering of the Standard Model.

3.6 AI–BASED MODEL BUILDING FOR DYNAMICAL SYSTEMS

With the leveraging of AI [1] and in particular, machine learning approaches [2, 3], dynamical systems have found a new fertile ground for further development [12, 13]. Showcase problems in applied mathematics, including the Lorenz strange attractor, reaction-diffusion spatio-temporal systems, and fluid dynamic flows captured by the Navier-Stokes equation, are being examined in a new guise as autoencoders identify parsimonious models with reduced dimensionality [14]. Such models are meant to provide the physical underpinnings of the phenomena enshrined in time-series data or generated by systems of raw differential equations.

Machine learning or more broadly, AI, is being leveraged for model discovery of dynamical systems underlying data represented as a time series. The data regression system, which in this context is a neural network predicting future behavior, is trained by the time series and regarded as the model itself. However, this model is heavily parametrized and hence too "fragile" to allow for extrapolation [12]. In other words, such models are not really amenable to yield physical laws, the way other data-regression approaches are [15, 21]. This statement has been voiced repeatedly and hints at some level of dissatisfaction: Machine learning, and AI in general, are very efficient at providing predictive models when trained on a sufficiently long time series but often do a poor job at providing physical insights regarding the underlying dynamical system.

This problem gets significantly amplified as we turn to biological or, more broadly, biomedical matter [19]. It is widely felt that, when examined in their multi-scale richness and complex heterogeneity, dynamical systems in biology or biomedicine cannot reach the level of maturation required to be subsumed into applied mathematics. This statement should be interpreted in the sense that we lack sparse enough models that provide physical underpinnings of biological/biomedical phenomena and are suitable for extrapolation. Will the leveraging of alternative AI–based approaches change the *status quo*? This chapter portends to address this problem and provide insights that will be methodologically fleshed out in the subsequent chapters to enable metamodel discovery by reverse engineering time series stemming from highly complex multi-scale realities.

A key question in fostering the mathematical maturation of model discovery in biology may be cast as follows: What constitutes a parsimonious model that provides physical underpinnings of biological phenomena? A standard answer with broad bearing on most problems considered tractable in applied mathematics is: "a sparse system of differential equations on a smooth manifold of latent coordinates" [14, 21]. As this chapter argues, this may be simply too much to ask for in the context of biological matter. Furthermore, this is not necessarily the format or framework that AI would typically choose for model discovery, given the "dimensionality curse" associated with the molecular reality of biological systems [19]. In principle, a single autoencoder that optimizes for sparsity in the discovery of the latent manifold might not provide a satisfactory solution to the modeling problem. Molecular reality *in vivo* typically has well over a million coordinates required to specify the state of the system, and the extent of connectivity parametrization for an autoencoder capable of handling such level of complexity would be simply enormous, implying that the training and variational optimization of the neural network would be off limits, at least with current computational capabilities [22].

This chapter squarely addresses this matter. To do so, it leverages AI methods to circumvent the difficulties associated with model discovery for time-dependent phenomena arising in soft or biological matter. In essence, as we shall show, AI dwells on a paradigm of what constitutes a parsimonious metamodel that is significantly different from the one adopted by applied mathematicians [19]. Thus, to identify the most economic yet faithful metamodel, AI will be shown to use two or more tandem autoencoders instead of one, as it is typically done in model discovery [14]. The autoencoders are coupled and become fully compatible with each other at the completion of the parameter optimization process, as defined precisely in the subsequent sections. In the simpler cases where two autoencoders are required, the second autoencoder translates the dynamic information embossed in the latent manifold, turning it into a topological dynamics metamodel [23] that can be decoded and enables significant propagation of the dynamics into the future. This property is essential for coverage of realistic timescales relevant to the level of state extrapolation required. Crucially, the topological dynamics metamodel constructed by AI is essentially a pattern-based model, not a system of differential equations, as it would be expected for standard model discovery in dynamical systems. This does not mean that AI is discovering laws without equations, but simply that AI adopts a different way of casting models susceptible of extrapolation as recognizable physics laws. Thus, AI may not straightforwardly give us the equations that govern *in vivo* protein folding, but the underlying physics discovered may in all likelihood be cast in terms of AI–interpretable topological patterns that signal the commitment of the chain to fold into a steady conformation [20].

These topological dynamics methods are likely to advance the field of AI–based model discovery as they enable the reverse engineering of time-series data stemming from vastly complex hierarchical realities. Such contexts are illustrated, for example, by the cellular setting that assists and expedites molecular processes, which have been hitherto considered off limits to machine learning approaches to dynamical systems.

At this stage of development, the topological methods introduced yield AI–recognizable patterns but do not beget latent differential equations that have traditionally been the hallmarks of model discovery. This may be viewed as a limitation in some sense, but we argue that that assertion is perhaps a reflection of narrow-mindedness. Synergistic efforts involving AI are likely to dominate future human endeavors in science, and AI systems are very much attuned to encode and process patterns in metamodels such as those produced by the topological approaches introduced in this book.

The impact of the topological methodology is likely to be broad, as it would render tractable problems in dynamical model discovery that have been hitherto considered off limits by applied mathematicians who are currently incorporating machine learning in their toolbox. Thus, ultra-complex realities recreating cellular environments that influence and steer molecular dynamics are likely to be within reach as topological methods are incorporated to the AI–empowered metamodel discovery.

3.7 AI–BASED SIMPLIFICATION OF A DYNAMICAL SYSTEM

Since the dawn of modern science, western civilization upheld the belief that understanding the workings of the universe pivots on finding fundamental equations that govern physical processes. In the case of a dynamical system, the underlying differential equations represent the basic model assumed to enshrine the physical laws that underpin the process described. The specific constraints, conservation principles, and symmetries of the system must all be taken into account when positing the differential equations. After centuries of work within this paradigm, it would be interesting to see how the leveraging of AI in synergy with the human endeavor will affect the choice of the format within which the physical laws are encoded and how this choice impacts the field of dynamic modeling. With the exponential development of computer technologies [22], we may soon be witnessing a paradigm revision. Equation-based modeling may or may not remain the dominant paradigm. Other types of dynamic descriptors are already making strides and contributing to a new understanding of the universe, or at least, of the dynamic multi-scale complexities of reality [19]. These descriptors are certainly different and arguably more pliable than what biological humans managed to achieve so far.

The discovery of physical laws distilled from sequential data representing a time series remains a major challenge as well as an imperative to enhance our understanding and control of physical processes. In recent times, this type of data-driven model discovery has been fueled by significant breakthroughs [15, 19, 21, 24]. Yet, we also live in an era of big data, especially stemming from fields like biology and astrophysics, where the multi-scale complexity of the data organization makes model discovery particularly daunting. It is unclear whether the enormous richness of the data describing *in vivo* contexts at the molecular level is amenable to the kind of parsimonious models based on differential equations that mankind holds dear and has sought since time immemorial. More than in any other field, in biology it is likely that a willy-nilly application of Occam's razor may lead to self-inflicted wounds.

Deep learning (DL) approaches realized through autoencoder architectures have proven particularly valuable for the discovery of data-driven models represented by differential equations framed on "essential coordinates." The latter, often referred to as "latent coordinates" [21], span the so-called center manifold in dynamical systems [16, 19]. This reduction entails a significant dimensionality reduction, and usually identifies the enslaving "slow" process that dynamically subordinates fast-relaxing modes [12, 16] and serves to encode the physical process we seek to model. Latent coordinates need to be selected very carefully, not only by weighing the extent of dimensionality reduction and the compactness and smoothness of the latent manifold, but also the economy of the resulting model. This economy is typically assessed by the complexity of the differential equations in latent space, quantified by the number of nonlinear terms [14].

In generic terms, an autoencoder seeks to identify latent coordinates $\mathbf{x}$ as output of a feed-forward neural network (NN) with multiple hidden layers. The NN is inputted observable vectors, denoted generically as $\mathbf{z}$, that serve to train the network, as the decoding of $\mathbf{z}$ from the latent

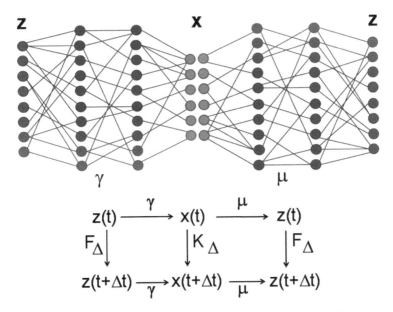

FIGURE 3.8 Generic scheme of the neural network (NN) architecture for autoencoder (γ, μ). The vector $z \in W$ represents the state of the system and $x \in \Omega$ represents an encoded latent state that is decoded back into z. The encoding process entails a dimensionality reduction with $dim\ \Omega < dimW$ and, concomitantly, a coarse graining of time in multiples of a time step Δ associated with propagation of the dynamics on the latent manifold Ω. Grey disks represent nodes in hidden layers for encoder $\gamma = \gamma(\Theta)$ and decoder $\mu = \mu(\Theta)$, where Θ denotes the weights of node connections that are variationally optimized according to a loss function $\mathcal{L} = \mathcal{L}(\Theta)$. The weights realizing the variational mimimum $(\Theta = \arg \min \mathcal{L})$ in the training of the NN are optimal at making the diagram commutative.

coordinates must reproduce the inputted z-value (Figure 3.8). The autoencoder is optimized variationally, meaning that the activation weight parametrization of the multilayered encoding and decoding functions, denoted respectively γ and μ, minimizes the loss function $\mathcal{L}(\gamma, \mu)$ that measures the efficacy in the recovery of the input vectors:

$$\mathcal{L}(\gamma, \mu) = Q^{-1} \sum_{z \in \mathfrak{F}} \| z - (\mu \circ \gamma) z \|^2; \gamma, \mu = \arg \min_{\gamma, \mu} \mathcal{L} \qquad (3.3)$$

where $\mathfrak{F}$ is the training set and $Q = |\mathfrak{F}|$ is its cardinal, for now regarded as a hyperparameter fixed to avoid overfitting [2, 3, 12, 13] in accord with the dimensions of the NN.

The autoencoder is functionally operative to model the dynamical system that underlies the given time series $\{z(t_0 + n \, \Delta \, t), z(t_0 + (n + 1) \, \Delta \, t)\}_{n = 0, 1, 2, ..., L}$ if and only if the following relations hold for $t = t_0 + n \, \Delta \, t$ with $n = 0,1,2,...,$ $L \, (L \gg 1)$ and any initial time t_0:

$$\left(K_\Delta \circ \gamma\right)z\left(t\right) = \gamma z\left(t + \Delta t\right) = \left(\gamma \circ F_\Delta\right)z\left(t\right);$$
$$\left(F_\Delta \circ \mu \circ \gamma\right)z\left(t\right) = \left(\mu \circ K_\Delta \circ \gamma\right)z\left(t\right) = z\left(t + \Delta t\right) \tag{3.4}$$

where F_Δ, K_Δ are the infinitesimal time maps in state space W and latent manifold Ω, respectively (cf. Figure 3.9). In other words, as the autoencoder becomes variationally optimized, the functional relations $K_\Delta \circ \gamma = \gamma \circ F_\Delta$; $F_\Delta \circ \mu = \mu \circ K_\Delta$ hold.

Thus, diagram commutativity (Figure 3.8) becomes the key property that enables us to assert that the choice of latent coordinates $x \in \Omega$ was the "right" one in the sense that it captures the entrainment of the dynamics on W by the reduced dynamics on the latent manifold Ω. The commutativity rules ensuring the compatibility of raw and latent dynamics may be cast in terms of derivatives using the chain rule [14] as Δt is taken to the infinitesimal limit dt.

3.8 MODEL DISCOVERY WITH DEEP LEARNING

Model discovery in dynamical systems has been maturing for some time, to the extent that it has been integrated into the corpus of applied mathematics. Furthermore, many applied mathematicians have incorporated DL to their toolbox as they make forays in reverse engineering of dynamical systems [12]. The overarching goal is to develop regression methods that enable extraction of parsimonious dynamics from large data organized as time series $\{z(t)\}_t$ with t given as multiples of a fixed time step [12, 14]. With the advent of DL, autoencoder architectures have successfully identified latent (essential) coordinate frames with significant dimensionality reduction (dimΩ < dimW). Ultimately, the parsimonious description should become solvable in the form of a differential system $\dot{x} = K\left(x\right)$, which spans a differential system $\dot{z} = F\left(z\right)$ for the time-series vector z. Yet, a mere stark reduction in dimensionality does not necessarily ensure a parsimonious description [14]. The variational optimization of the autoencoder should simultaneously consider both the dimensionality of the coordinate frame for the latent manifold and the sparsity of the equations in the latent manifold, with the latter measured by the number of terms in the flow (vector field) that determines $\dot{z}$.

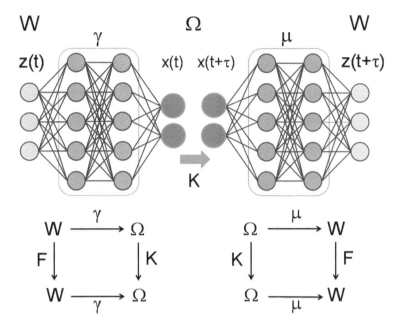

FIGURE 3.9 Scheme of autoencoder for latent dynamics optimizing simultaneously the dimensionality reduction and the parsimony of latent dynamical system governed by the flow K. The optimization is extended further relative to the architecture displayed in Figure 3.8 by adding the extra term $\mathcal{L}_s(K)$ to yield the loss functional $\mathcal{L} = \mathcal{L}(\Theta, K) = \mathcal{L}(\Theta) + \mathcal{L}_s(K)$. The optimal Θ, K pair realizing the variational minimum is precisely the one that makes the bottom diagrams commutative.

The autoencoder architecture and the required dual diagram commutativity between the latent flow K and the given time series flow F are jointly represented in Figure 3.9. We adopt a time step t as the time-series hyperparameter to discretize the flows by coarse graining time resolution. The loss function $\mathcal{L}(\gamma, \mu, K)$ now includes five terms: $\mathcal{L}_r(\gamma, \mu)$, previously introduced, weights the efficacy of the recovery of the z-value upon encoding and decoding; $\mathcal{L}_s(K)$ accounts for the sparsity of the latent flow K; and the terms $\mathcal{L}_C(\gamma, K), \mathcal{L}_C(\gamma, \mu), \mathcal{L}_C(\gamma, \mu, K)$ represent the penalties associated with imperfect diagram commutativity that need to be imposed to guarantee the dynamic compatibility of the z and x-descriptions of the system. Thus, we get

$$\mathcal{L}(\gamma, \mu, K) = \mathcal{L}_r(\gamma, \mu) + \vartheta_s \mathcal{L}_s(K) + \vartheta_C \left[\mathcal{L}_C(\gamma, K) + \mathcal{L}_C(\gamma, \mu) + \mathcal{L}_C(\gamma, \mu, K) \right],$$

$$(3.5)$$

where the relative weights ϑ_s, ϑ_C are hyperparameters and

$$\mathcal{L}_r\left(\gamma,\mu\right)=Q^{-1}\sum_{z\in\mathfrak{F}}\left\|z-\left(\mu\circ\gamma\right)z\right\|^2, \tag{3.6}$$

$$\mathcal{L}_C\left(\gamma,K\right)=Q^{-1}\sum_{z\in\mathfrak{F}}\left\|\left(K\circ\gamma\right)z-\left(\gamma\circ F\right)z\right\|^2, \tag{3.7}$$

$$\mathcal{L}_C\left(\gamma,\mu\right)=Q^{-1}\sum_{z\in\mathfrak{F}}\left\|\left(F\circ\mu\circ\gamma\right)z-\left(\mu\circ\gamma\circ F\right)z\right\|^2, \tag{3.8}$$

$$\mathcal{L}_C\left(\gamma,\mu,K\right)=Q^{-1}\sum_{z\in\mathfrak{F}}\left\|\left(\mu\circ K\circ\gamma\right)z-\left(F\circ\mu\circ\gamma\right)z\right\|^2, \tag{3.9}$$

with $Q=\left|\mathfrak{F}\right|$ indicating the cardinal of the training set.

To determine the sparsity term $\mathcal{L}_s\left(K\right)$, we first take the limit case where t becomes the time infinitesimal dt and assume that $\dot{x}=K\left(x\right)=AP\left(x\right)$, where $A=[Aij]$ is a $m\times p$ matrix ($m=\dim\Omega$) whose coefficients are variationally optimized jointly with the parametrization of the autoencoder maps γ and μ, and $P(x)$ is a p-vector given by:

$$P\left(x\right)^T=\left(1\ x_1\ x_2\ x_3\ x_1^2\ x_2^2\ x_3^2\ x_1x_2\ x_2x_3\ x_1x_3\ x_1x_2x_3...\right), \tag{3.10}$$

consisting of p functional terms. The terms do not need be of polynomial form, and they may be chosen in accord with the type of symbolic regression. Thus, the variational optimization of A is tantamount to a symbolic regression in the latent manifold Ω. To ensure the sparsity of the model we define the loss term as $\mathcal{L}_s\left(K\right)=\|A\|_F$, where $\|.\|_F$ is the Frobenius norm given by

$$\|A\|_F=\left[\sum_{i=1}^{m}\sum_{j=1}^{p}A_{ij}^2\right]^{1/2}. \tag{3.11}$$

The incorporation of this term into the loss function given by Equation (3.5) ensures that the latent manifold is selected to provide a parsimonious model with reduced dimensionality governed by the simplest possible set of differential equations in consonance with the time series provided. A different but equivalent formulation of the autoencoder has been given elsewhere [14] and uses the chain rule of derivation rather than diagram commutativity to infer the proper loss functional.

3.9 AI EMPOWERS MOLECULAR DYNAMICS

As applied mathematicians make their forays combining dynamical systems with machine learning, they adopt specific systems that have traditionally represented showcases for model building. Examples of such systems are the Lorenz attractor generated by three coupled ordinary differential equations, the spatio-temporal reaction-diffusion systems, and the hydrodynamics and turbulence models governed by the Navier-Stokes equation [14, 24]. Seldom, if ever, have we had a chance to evaluate how autoencoder techniques pan out in the realm of molecular dynamics for ultra-complex multiscale many-body systems. We are specifically referring to the discovery of models that distill the cooperative collective motion of ensembles of atoms, atomic groups, molecules, or molecular assemblages in condensed phases characterized as clusters, liquids, glasses, crystals, polymers, and so forth, under overarching categories such as soft matter and biological matter. The foundational approach to model discovery for such systems has been traditionally provided by statistical mechanics. These methods have met considerable success, except in the realm of biological matter, where problems such as the discovery of protein-folding pathways, or the role of *in vivo/* cellular contexts in expediting the folding process cannot be even remotely addressed due to their sheer complexity and heterogeneity. Interacting molecular units are simply too diverse and the cellular environment is too complex at multiple scales and heterogeneous for statistical mechanics to make successful forays in biology [19, 25, 26]. This is precisely the context where AI may empower dynamical systems by leveraging highly specialized autoencoders and even batteries of autoencoders, as described subsequently.

To furnish a general framework, we may start by noting that we are dealing with N particles that may be free or tethered through covalent linkages forming assemblages such as biopolymers that may interact with one other through ephemeral or permanent associations that do not involve covalent bonds. A protein chain embedded in an aqueous environment and interacting with other biomolecular entities such as protein enzymes, chaperones, or ribosomes represents the quintessential situation that we wish to address in this book as we learn to leverage AI methods and integrate them on a model discovery platform.

Keeping for now the discussion at its most generic level, let us consider an ensemble of N particles that has associated with it an internal energy $U = U(\hat{z}), \hat{z} \in \mathbb{R}^{3N}$ only dependent on inter-particle distances and hence invariant upon isometries – distance-preserving maps – of $\mathbb{R}^3$. Then, the state or configuration of the system may be represented as a point z in the quotient space $W = \mathbb{R}^{3N}/E^+(3)$, where $E^+(3)$ is the special Euclidean group of isometries of $\mathbb{R}^3$ including only rigid-body translations and rotations but excluding reflections [27]. The latter are excluded because they do not preserve the chirality (handedness) of asymmetric tetrahedral carbon groups, which constitutes a key constraint: Chirality is known to be strictly preserved in biology. Thus, W is (3N-6)-dimensional and its coordinates represent all the internal degrees of freedom of the system. In addition to the reduction modulo isometries, another quotient is required to represent the latent manifold Ω. This further dimensionality reduction depends on the constraints brought about by the covalent linkages that tether specific units in the system. For example, if we are studying the dynamics of protein folding, we note that high-frequency vibrational motions involving covalently paired atoms may be averaged out as their associated timescales, in the femtosecond (fs) range [28], are incommensurably shorter than those associated with soft modes represented by dihedral torsions of the polymer backbone (Figure 3.10), typically in the nanosecond (ns) or sub-ns range. A similar reduction applies to planar angular vibrations, with frequencies in the order of $(fs)^{-1}$ to $(10fs)^{-1}$ [19, 28]. These simplifications point to a parsimonious model representing an adiabatic system that incorporates only soft modes of the chain. Thus, as protein folding is represented as a dynamical system with the protein chain searching in conformation space in an *in-vitro* setting, Ω becomes a compact manifold in the form of a 2M-torus, where M is the number of amino acid units in the chain [20]. This latent manifold is deduced assuming that the water molecules surrounding the protein chain that explores conformation space are treated implicitly, that is, their influence is energetically subsumed in the potential energy function U only dependent on distances between the protein chain subunits and the local configurations that result thereof.

The dynamics in W obeys the basic physical law:

$$\dot{z} = v; \dot{v} = -\nabla_z U. \tag{3.12}$$

To properly define $U = U(z)$, we introduce the following

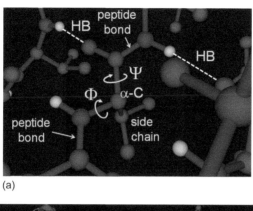

(a)

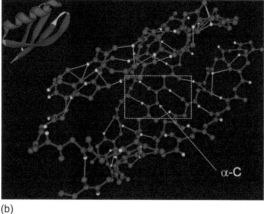

(b)

FIGURE 3.10 Latent coordinates for a folding protein chain. (**a**) Individual residue unit within a protein chain attached to an adjacent residue through a torsionally rigid linkage known as a peptide bond. For the sake of illustration, a particular protein has been selected, namely the thermophilic variant of the B1 domain of protein G from *Streptococcus*, whose native 3D-structure is found at the entry 1GB4 in the protein data bank. In the latent coordinate frame, the state of each residue is represented by a pair of coordinates (Φ, Ψ) representing the torsional dihedral degrees of freedom of the protein backbone. The amino acid type for the residue, describing the local chemical composition of the chain, is identified by the side-chain group that is covalently linked to the alpha-carbon in the protein backbone. In turn, polar groups in the backbone and side chain can be engaged in orientationally and distance-dependent noncovalent linkages known as hydrogen bonds (HB, dashed lines), whose cumulative folding-stabilizing effect becomes a determinant of the protein 3D structure. (**b**) Zoom out of the detail shown in (a), revealing the backbone HB pattern and topology (ribbon rendering) of the native structure. Backbone HBs exposed to water are indicated by thin green lines (buried ones in grey) and they signal structural deficiencies, since the structure may get locally disrupted when the polar groups paired by intramolecular HBs get hydrated, that is, interact with surrounding molecules of the aqueous solvent.

Theorem 3.1

There exists a map $U : \mathbb{R}^{3N}/E^{+}(3) \rightarrow \mathbb{R}$ that makes the following diagram commutative:

$$
\begin{array}{c}
\mathcal{U} \\
\mathbb{R}^{3N} \rightarrow \mathbb{R} \\
\pi\downarrow \quad \nearrow U \\
\mathbb{R}^{3N}/E^{+}(3)
\end{array}
\qquad \Rightarrow \qquad U \circ \pi = \mathcal{U}
\tag{3.13}
$$

■

Where $\pi : \mathbb{R}^{3N} \rightarrow \mathbb{R}^{3N}/E^{+}(3)$ is the canonical projection associating each point in $\mathbb{R}^{3N}$, specifying the 3D-coordinates of each of the N particles of the system, with the set of points in $\mathbb{R}^{3N}$ resulting from rotations and translations of the N-particle system treated as a rigid body, that is, preserving all inter-particle distances. Thus, the projection associates a state of the system with its class in quotient space, that is, with the collection of all points contained in the group orbit generated by the action of the Euclidean group on the point in the domain.

To prove this "factorization" result, it suffices to note that the potential energy $\mathcal{U} : \mathbb{R}^{3N} \rightarrow \mathbb{R}, \mathcal{U} = \mathcal{U}(\hat{z}), \hat{z} \in \mathbb{R}^{3N}$ is invariant on the orbits of $E^{+}(3)$ in $\mathbb{R}^{3N}$.

An autoencoder (γ, μ) yielding a parsimonious model in Ω would require that the autoencoder constructs $\tilde{U}(x) = U(\mu(x))$, so the dynamic equations in the latent manifold become

$$
\dot{x} = \tilde{v}; \dot{\tilde{v}} = -\nabla_x \tilde{U}(x). \tag{3.14}
$$

This implies that in the infinitesimal limit $\tau = dt$, the potential energy function may be obtained by noting that the sparse map K defined by the autoencoder obeys

$$
\dot{x}(t) = K(x(t)) = -\int_0^t \nabla_x \tilde{U}(x(t'))dt' + \dot{x}(0). \tag{3.15}
$$

On the other hand, if the time series used to train the encoder is generated using molecular dynamics governed by potential energy function $U = U(z)$, then an additional loss term in $\dot{x}$ of the form

$$
\mathcal{L}_U(\mu, K) = \left\| K(x(t)) - \nabla_z x(t) \left[-\int_0^t \nabla_z U(z(t'))dt' + \dot{z}(0) \right]_{z(t') = \mu(x(t'))} \right\|^2
\tag{3.16}
$$

needs to be incorporated to optimize the autoencoder and its associated propagator **K**.

3.10 PHYSICS ON LATENT MANIFOLDS

The sheer complexity of the molecular dynamics arising in soft and biological matter, especially in *in vivo* settings where N~10^6–10^7 including solvent molecules, is unlikely to ultimately allow for the type of reductive approach that standard autoencoders usually provide (Figure 3.9). A case in point is the discovery of the physical underpinnings of protein folding assisted by an *in vivo* context that enhances the expediency of the process (Figure 3.11) [29]. The space is often anisotropic, the system itself is highly heterogeneous, and its components are too diverse, with

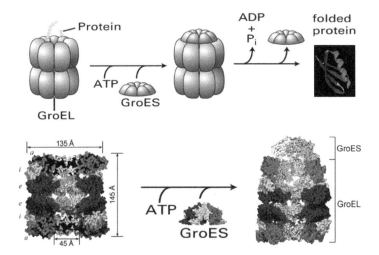

FIGURE 3.11 *In vivo* setting assisting the folding of a protein chain. A molecular cage known as chaperone GroEL assists the folding process, enhacing its expediency well above the level of efficiency that may be achieved *in vitro*, that is, in the test tube. The cage consists of a dimer of two annular molecular assemblages, each consisting of a complex of seven identical proteins. Each protein is made of three regions known as apical, intermediate, and equatorial domain, denoted respectively "a, i, e" in the figure. These domains undergo a certain amount of conformational rearrangement upon binding to the cell-fuel molecule ATP (adenine triphosphate). This rearrangement in turn enables the cage dimer to incorporate a molecular lid, known as GroES. Once the lid is on, the protein inside the cage is subject to a number of iterative annealing steps through interactions with the proteins lining the interior of the cage. This annealing process enables the protein to avoid getting kinetically trapped in misfolded states. Upon release of ATP, the lid becomes detached and the protein exits the system regardless of whether it has satisfactorily completed the folding process or not. If the latter is the case, additional catalytic cycles engaging the same or other folding assistants may be necessary.

potentials or force fields that cannot fully account for the complexities and many-body effects enshrined in the time series. It is doubtful that the latent compact manifold spanning soft-mode coordinates (backbone torsional dihedrals in the case of the folding protein, see Figure 3.10) will be amenable to the type of model discovery that is usually cast in terms of sparse differential equations. For example, generating a minimal set of 2M (M~100) coupled differential equations that govern the backbone torsional dynamics underlying the protein-folding process with implicit treatment of the environment is out of reach given current capabilities in deep learning.

Other many-body systems share similar problems, as their wanton complexity is off limits for state-of-the-art autoencoders seeking to identify parsimonious models with differential equations. Yet, as we shall now show, a generic topological understanding of the latent dynamics may yield a way of learning dynamic data that enables suitable propagation of the time series into the future, endowing the autoencoder with predictive value. Thus, topological methods will be readily incorporated in AI–based model discovery for systems with multi-scale complexity.

The approach entails a simplification based on the topological dynamics in the latent manifold, that is, on the dynamics modulo the basins of attraction of the generic singular points of the map $K : \Omega \to T\Omega$, where $T\Omega$ denotes the tangent bundle of Ω and Ω itself is assumed to be C^1-differentiable and compact [30], as it is the case in the previously discussed example.

To make further progress with the argument we first prove the following result:

Theorem 3.2

Under the assumptions of Section 3.9, the autoencoded map K yields no closed orbits in Ω

By *reductio ad absurdum* let us assume $x(0) = x(T)$ for $T > 0$. Then, we get

$$0 = \int_0^T \tilde{U}\big(x(t)\big)dt = \oint d\left[\int_0^t \tilde{U}\big(x(t')\big)dt'\right] = \int_0^T\left[\nabla_x \int_0^t \tilde{U}\big(x(t')\big)dt'\right]\dot{x}(t)dt$$

$$= \int_0^T\left[\int_0^t \nabla_x\tilde{U}\big(x(t')\big)dt'\right]\dot{x}(t)dt = -\int_0^T \|\tilde{v}\|^2\,dt \le 0. \tag{3.17}$$

Equation (3.17) implies that $\tilde{v} \equiv 0$, which is absurd since $x(0)$ was not assumed to be a steady state but a point in a closed orbit. Q.E.D. ▪

Given that Ω is a compact manifold, this result has far-reaching consequences [30, 31]:

1) The latent dynamics have no attractors made up of recurrent orbits, and

2) Since there is no circulation around them, all singular points of the latent flow are hyperbolic, hence generic, since the real part of the eigenvalues of the Jacobian at the singular points cannot be zero.

This implies the following result:

Corollary 3.1

The latent dynamics governed by Equation (3.13) are of the Morse-Smale type [31], hence structurally stable, that is, qualitatively (topologically) invariant under small perturbations. ▪

To rigorously define structural stability, we first note that the space $\mathfrak{H}(\Omega)$ of smooth maps $\Omega \to T\Omega$ is endowed with a natural metric inherited from the supremum norm given by

$$
\begin{aligned}
\|H\|_{\text{sup}} &= Sup_{x\in\Omega,\, j=1,\dots,\dim\Omega}\left[\left\|H(x)\right\|, \left|\frac{\partial}{\partial x_j}H(x)\right|\right] \\
&= \text{Max}_{x\in\Omega,\, j=1,\dots,\dim\Omega}\left[\left\|H(x)\right\|, \left|\frac{\partial}{\partial x_j}H(x)\right|\right]
\end{aligned}
\tag{3.18}
$$

for $H\epsilon\mathfrak{H}(\Omega)$. Then, to state that the latent flow $K(x)$ is structurally stable means that for any given Δ-neighborhood of K, $\mathfrak{B}_\Delta(K) \subset \mathfrak{H}(\Omega)$, there exists a value $\varepsilon=\varepsilon(\Delta)$, such that for any $G \in \mathfrak{B}_\Delta(K)$ there exists an ε-homeomorphism $h_\varepsilon : \Omega \to \Omega$ satisfying $\max_{x\in\Omega}|x - h_\varepsilon(x)| < \varepsilon$ that transforms trajectories of K onto trajectories of G [31].

Given the qualitative invariance of the latent flow under small perturbations, the following observation is key to justify the leverage of AI to construct dynamic models based on time series: *The structural stability of the latent dynamics is essential to enable model discovery in view of the fact that the exact parameters determining the potential energy U of the many-body system are not known precisely.*

Given the characterization of the latent flow given by theorem 3.2 and its corollary 3.1, we may encode the flow in a simplified manner, as we now build a metamodel. To that effect, we first define the equivalence relation "~" for any pair $x,y \in \Omega : x{\sim}y \Leftrightarrow \omega(x) = \omega(y)$, where $\omega(x)$ denotes destiny (omega) state given by the limit at $t \to \infty$ of the trajectory initiated at x [32]. Since the singular points are the steady states of the system, the latent flow may be encoded by the equivalence classes identified as the basins of attraction of the singular points. As points in Ω are regarded "modulo basins," we have in effect defined the quotient space $\Omega/{\sim}$ as the set of basins of attraction and separatrices (basins of lower dimension) of critical generic points that partition the manifold. The quotient space is relatively simple to encode since the singular points of the latent flow are isolated and finite in number. To demonstrate this proposition, we note that otherwise, if they were infinite in number, they would have an accumulation point since the latent manifold is compact, and that cannot happen because all singular points are generic.

Let us then denote by $\Gamma : \Omega/{\sim} \to \Omega/{\sim}$ the coarse-grained flow that determines the inter-basin transitions, while $\pi : \Omega \to \Omega/{\sim}$ denotes the canonical projection that associates a point in the latent manifold with its equivalence class. The flow must be such that the diagram in Figure 3.12 becomes commutative, and specifically, the following flow-compatibility relations must hold:

$$\mu^{\#} \circ \pi = \mu, \tag{3.19}$$

$$\Gamma \circ \pi = \pi \circ K, \tag{3.20}$$

$$\mu^{\#} \circ \Gamma = F \circ \mu^{\#}. \tag{3.21}$$

Thus, to parametrize Γ, we need to introduce a second autoencoder with variational parameter optimization determined by the loss function

$$\mathcal{L}_{\sim}\left(\mu^{\#},\Gamma\right) = \mathcal{L}_{\mu}\left(\mu^{\#}\right) + \mathcal{L}_{\sim}\left(\Gamma\right) + \mathcal{L}_{\sim}\left(\mu^{\#}\right), \tag{3.22}$$

where:

$$\mathcal{L}_{\mu}\left(\mu^{\#}\right) = Q^{-1}\sum_{x\in\mathfrak{F}}\left\|\left(\mu^{\#}\circ\pi - \mu\right)x\right\|^{2}, \tag{3.23}$$

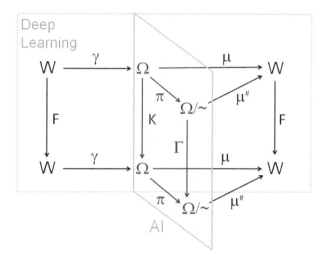

FIGURE 3.12 Scheme of a metamodel consisting of two coupled variational autoencoders (γ, μ), $(\pi, \mu^{\#})$ required to generate the discrete flow Γ that propagates the topological dynamics in the latent quotient manifold $\Omega/\sim$. The parameter optimization of both autoencoders ensures the full commutativity of the diagram. The discovery of the topological metamodel $\Gamma : \Omega/\sim \rightarrow \Omega/\sim$ hinges on a CNN–based construction of the modulo-basin projection $\pi : \Omega \rightarrow \Omega/\sim$ of the latent dynamics. This scheme introduces a level of coarse graining that is more drastic than that adopted by conventional autoencoders for model discovery of time-series data.

$$\mathcal{L}_{\sim}\left(\Gamma\right) = Q^{-1} \sum_{x \in \mathfrak{F}} \left\| \left(\Gamma \circ \pi - \pi \circ K\right) x \right\|^{2}, \qquad (3.24)$$

$$\mathcal{L}_{\sim}\left(\mu^{\#}\right) = Q^{-1} \sum_{x \in \mathfrak{F}} \left\| \left(\mu^{\#} \circ \Gamma - F \circ \mu^{\#}\right) \pi x \right\|^{2}, \qquad (3.25)$$

where the training set is denoted $\mathfrak{F} \subset U$, with $Q = |\mathfrak{F}|$ and $\mu^{\#}$ denotes the decoder for the quotient space.

The commutativity of the diagram in Figure 3.12 implies that the ultimate simplicity in a model governing ultra-complex many-body problems, such as identifying *in vivo* protein-folding trajectories, may be achieved by projecting the latent dynamics onto the quotient manifold $\Omega/\sim$.

3.11 DYNAMICS IN QUOTIENT SPACE

The sparse latent dynamics obtained by leveraging autoencoders that serve as model discoverers has become the subject of intense research in applied mathematics. Such methods are less suited to unravel underlying laws in dynamical systems that represent biological or soft matter, where the number of internal degrees of freedom is astronomical. This book proposes to couple two commutative – hence compatible – autoencoders as described in Figure 3.12, yielding a factorization of the latent dynamics through the quotient space $\Omega/\sim$. The commutativity of the whole diagram displayed in Figure 3.12 ensures the compatibility of the different levels of coarse graining of the dynamics that constitute the metamodel. In turn, the diagram commutativity (Eqs. (3.19)–(3.21)) is subsumed into the variational functional of the autoencoder (Eqs. (3.22)–(3.25)) so that optimization of the underlying neural networks is equated with – or rather becomes as close as possible to – diagram commutativity. Identifying the quotient space $\Omega/\sim$ under the "modulo basin" equivalence relation defined in the previous section is akin to a pattern recognition process, where time series datapoints are plotted onto 2D cross sections of the latent manifold Ω. The task may be entrusted to a CNN and becomes enormously simplified as corollary 3.1, jointly with theorem 3.2, guarantee a finite number of basins of attraction for isolated singular points that are all generic, that is, topologically equivalent under small perturbations of the latent vector field.

For example, in the case of the folding protein, the 2D cross section is the (Φ,Ψ)-torus (Figure 3.10a), and a typical time series for latent dynamics governed by Equation (3.14) is given in Figure 3.13 [19, 20]. The CNN "discovers" the topological metamodel which can be represented as a graph with vertices corresponding to the basins of attraction of the generic minima and edges connecting basins of attraction (Figure 3.13) in a manner specified by the autoencoder-generated information fed onto the CNN. More specifically, the graph is generated in accord with the inferred topography of the 2-torus cross section of the potential energy map $\tilde{U}:\Omega \to \mathbb{R}$, in turn generated by the autoencoder.

To generate the entire quotient space within a graph representation we need to integrate the cross sections corresponding to the 2D-projections of $\Omega/\sim$. As an illustration, let us consider a GG dipeptide ($M = 2$, $G = $ glycine). The latent manifold is a Cartesian product of two tori (Figure 3.14), one for each pair of (Φ,Ψ)-dihedrals coordinates specifying the backbone conformation of the respective residue [20]. The topological representation of the vector field steering the backbone torsional dynamics of a generic residue in the protein chain is given in Figure 3.13. Opposite sides of the square are identified as per the $\pm180°$ identification of

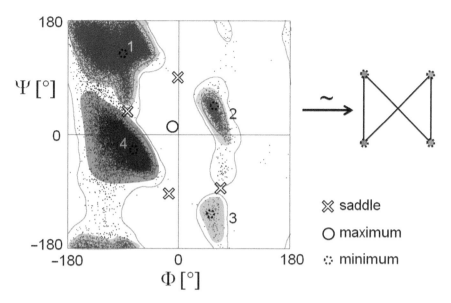

FIGURE 3.13 Toroidal (Φ,Ψ) cross section of the time-series data for an individual unit along a folding protein chain. The associated cross section of the dynamical system representing the protein folding process is discovered through pattern recognition leveraging a CNN, and its modulo-basin representation in the cross section of the latent quotient manifold is given as a graph. The vertices in the graph are critical points corresponding to omega-sets for all points in their respective basins of attraction, while the edges indicate allowed inter-basin transitions that determine the topological dynamics.

(Φ,Ψ)-dihedrals determining the local torsional state of the backbone. The four colored sectors morph topologically into the allowed valleys in potential energy. The organization of the basins of critical points is compatible with the underlying 2-torus and topologically represented by a graph (Figure 3.13) for each residue. The bottom panel in Figure 3.14 represents the quotient space for the dipeptide chain [20]. For each residue, the quotient space cross section is represented by a graph with vertices indicating 2-dimensional basins, and an edge linking two vertices indicates that the respective basins are connected through a line of steepest descent crossing a saddle point and orthogonal to the separatrix at the saddle. For a protein chain consisting of two consecutive residues, denoted 1 and 2, the quotient space becomes the tensor product of the two graphs where vertices now denote basin pairs (B_1, B_2), and where basin pair (B_1, B_2) is connected with (B'_1, B'_2) if and only if B_1 is connected with B'_1 or $B_1 = B'_1$, and B_2 is connected with B'_2 or $B_2 = B'_2$. Thus, the quotient space for two residues

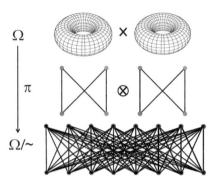

FIGURE 3.14   Reconstruction of the latent quotient manifold. The iterative process is represented by progressively incorporating adjacent-residue cross sections of the latent manifold as Cartesian products and, in parallel, reconstructing the quotient manifold as tensorial product of the individual graphs representing the cross sections of the quotient manifold (Figure 3.13). Within this representational framework, the topological dynamics becomes a walk in the quotient manifold graph.

consists of $4 \times 4 = 16$ vertices, where each vertex connects via one edge with eight other vertices and connects via two adjacent edges to the seven remaining vertices. Given the symmetry of the problem, to prove the assertion it suffices to note that the vertex denoting basin pair (1,1) is directly connected to eight other vertices denoting pairs (2,1), (4,1), (1,2), (1,4), (2,2), (2,4), (4,2), (4,4) while vertex (1,1) connects to all the remaining seven vertices-pairs containing basin 3 via two adjacent edges.

Thus, the projection on quotient space $\Omega/\sim$ of a latent MD trajectory on the 4-torus Ω becomes a walk on the tensorial product graph at the bottom of Figure 3.14 [20].

3.12 ENCODING HIERARCHICAL SYSTEMS WITH AI

Kurzweil and others have successfully argued that a hierarchical structure of reality is necessary for proper encoding and processing in a suitable AI–based inferential framework [22, 33]. To cast the discussion in the broadest terms, we shall refer to hierarchical as the attribute of a system endowed with nested complexities at multiple scales arising at different levels of description and providing different levels of coarse graining. This notion was delineated previously in this chapter, where a number of illustrations reveal the dynamic entrainment of fast-relaxing modes by slower modes spanning a latent manifold. Thus, just like in the adiabatic approximation [16], fast motions are averaged out and hence treated implicitly in a coarse-grained version of the dynamics focusing on longer timescales, with an autoencoder providing the inferential framework to discover the latent coordinate system.

Indeed, an escalation in the level of coarse graining, from the subatomic to the atomic, molecular, subcellular, and beyond, has always suggested that the hierarchical structure of reality may span over several layers, with nested complexities where information at a peripheral layer is incorporated implicitly at a core layer. In topological terms, we envision a whole sequence of quotient manifolds as we lump up states at different levels of description within progressively coarser equivalence classes, with dynamic compatibility of the various descriptions imposed by commutative flow diagrams (cf. Figure 3.15). Thus, in the broadest sense, an encoding of a hierarchical dynamical system within an AI–based inferential framework may require a battery of tandem autoencoders whose dynamic compatibility in an optimized parametrization is ensured by the commutativity of the diagram that combines the different levels of flow encoding. Figure 3.15 shows one such diagram reflecting the dynamic interplay of three autoencoders; one to generate the latent manifold Ω of intrinsic coordinates, one for a first-level quotient space $\Omega/\sim$, and one for a second-level (coarser) quotient space $\Omega/\sim/\approx$. The autoencoders are variationally optimized so

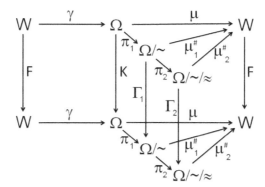

FIGURE 3.15 Metamodels for hierarchical dynamical systems. Autoencoders in tandem implemented to forecast dynamics that allow for a hierarchical description defined by two equivalence relations "$\sim$" and "$\approx$". Two autoencoders, $\{\pi_1, \mu^{\#}_1, \Gamma_1\}$ and $\{\pi_2, \mu^{\#}_2, \Gamma_2\}$, operating sequentially as required for topological encoding of the dynamics at the coarsest level in the latent quotient manifold $\Omega/\sim/\approx$. The F-compatibility of the latent flows K, Γ_1, Γ_2 in Ω, $\Omega/\sim$, $\Omega/\sim/\approx$, respectively, requires the full commutativity of the diagram. This commutativity is achieved by variational parameter optimization for the three autoencoders $\{\gamma, \mu, K\}$, $\{\pi_1, \mu^{\#}_1, \Gamma_1\}$, and $\{\pi_2, \mu^{\#}_2, \Gamma_2\}$.

that the flow diagram becomes commutative, which in turn reflects the compatibility of the different levels of dynamic encoding within the latent three-level hierarchy.

The subsequent chapters illustrate the implementation of tandem autoencoder technology to treat implicitly the hierarchical structure of the molecular dynamics of biological and biomedical processes. The relationship between hierarchical structure and tandem autoencoders of nested coarse graining descriptions is of universal applicability and may be extended to any number of equivalence relations, leading to progressively simplified metamodels.

3.13 QUANTUM AUTOENCODING OF GRAVITY

A theory of quantum gravity (QG), that is, a theory of gravity in accord with the tenets of quantum mechanics (QM), is often regarded as off-limits due to the major disparity in the dimensional scales where quantum and gravitational forces materialize [34, 35]. So far, gravity has not been quantized and hence a sense of incompleteness pervades the field. The unification of all forces of nature remains a holy grail in physics ever since Einstein's pursuit.

A different sort of grand unification was pursued by Albert Einstein. Since weak and strong nuclear forces were not clearly understood at the time, he sought to do it without involving QM, a theory he never fully endorsed. At first glance, quantum gravity stands almost as an oxymoron: After all, QM deals with the atomic and subatomic scales, while the best theory of gravity to date is Einstein's general relativity (GR), which is essentially classic, that is, non quantum, and deals mainly with cosmological scales (except for singularities). Einstein's theory of relativity postulates that high concentrations of energy and matter impinge on the curvature of space-time, deflecting the trajectories of particles, as it occurs in a gravitational field. This theory withstood admirably the long-term attrition of experimental corroboration. Yet, if we attempt to cast GR in QM terms, we need to deal with the fact that matter and space-time become "protean" at scales of the order of Planck's length (10^{-33}cm), akin to the sea of virtual particles that fill up empty space. In this essentially quantum world, the equations of GR no longer hold.

Pursuing QG still makes good sense when we deal with GR singularities, for example, black holes or the first few sub-attoseconds after the big bang, when the vast differences in material scales that set apart QM and GR can be reconciled, prior to the inflationary phase. Thus, in the spirit of this book, we need to address the question: *What would constitute a quantum metamodel of a theoretical model of gravity and how to construct it?*

This question was effectively formulated by the Argentinian physicist Juan Maldacena [34]. He focused on the anti de Sitter (AdS) space, a hyperbolic space that shares curvature properties with the sphere representing the event horizon of a black hole. Maldacena postulated that a string theory (ST) of gravity in a 5-dimensional AdS space ($AdS_5 = W$ in standard notation) is *equivalent* to a quantum field theory (QFT) on its boundary ∂W, which constitutes a 4-dimensional Minkowski space, akin to the one Einstein adopted for GR.

How could the boundary represent the latent space for the ST in an AdS? The answer is provided by the computation of the Shannon entropy of the black hole that assesses its total information storage capacity contained in all degrees of freedom [36]. These "ultimate" degrees of freedom involve, of course, atomic and subatomic entities, all the way to quarks and gluons, and ultimately those entities from hitherto unknown depths in the physical structure of matter, the so-called "level X." This "ultimate entropy" is proportional to the surface area of the event horizon [36].

In general, the boundary M_d of AdS_{d+1} is a d-dimensional Minkowski space with the symmetry group $SO(2, d)$ of AdS_{d+1} acting on M_d as the conformal (i.e., inner product-preserving) group. Thus, there are two ways to get a physical theory with $SO(2,d)$ symmetry: A relativistic field theory on AdS_{d+1} and a conformal field theory (CFT) on M_d. A suitable theory on AdS_{d+1} has been conjectured by Maldacena to be *equivalent* to a CFT on M_d [34]. The computation of observables of the CFT in terms of supergravity on AdS_{d+1} can and should be attempted using the methods described in this book. In accord with the tenets of topological metamodel autoencoding, M_d should be identified with the latent manifold Ω and the CFT on M_d "holographically" spanned onto AdS_{d+1} should be generated by an holographic autoencoder that exploits the $SO(2, d)$ to generate jointly the latent manifold together with its parsimonious metamodel.

Correlation functions in conformal field theory are given by the dependence of the supergravity action on the asymptotic behavior at infinity [35]. Thus, dimensions of operators in CFT are determined by masses of particles in string theory. It is thus conjectured that to describe the Yang-Mills theory in four dimensions, one should use the whole infinite tower of massive Kaluza-Klein states on $AdS_5 \times \mathbf{S}^5$. Chiral fields in the 4-dimensional $N = 4$ theory correspond to Kaluza-Klein harmonics on $AdS_5 \times \mathbf{S}^5$.

The spectrum of Kaluza-Klein excitations of $AdS_5 \times S^5$ are matched against operators of the $N = 4$ theory.

To discover the topological metamodel for superstring theory on $W = AdS_{d+1} \times S^{d+1}$, we first note that the boundary is topologically identified as $\partial W = (S^1 \times S^{d-1})/\mathbb{Z}_2$, where the group $\mathbb{Z}_2$ acts by rotation in π on S^1 and multiplication by -1 on S^{d-1}. In other words, the latent manifold fulfills the compactness condition, and to simplify the computation and straighten the symmetry we may use its universal cover [37], as demonstrated by Ariel Fernández and Oktay Sinanoglu: $\Omega \approx S^{d-1} \times \mathbb{R}$, with the real axis representing the time dimension.

Thus, to prove using AI the conjectured equivalence between $N = 4$ QFT on $\Omega \approx S^{d-1} \times \mathbb{R} = \partial W$ and Type IIB supergravity as string theory (ST) on $W = \langle AdS_5 \times S^5 \rangle$ ($\langle . \rangle$=universal cover [37]), we need a holographic autoencoder for a DL neural network capable to represent the ST on the space W. This DL system has been constructed [38]. The autoencoder should identify the latent manifold as $\Omega \approx S^3 \times \mathbb{R}$ taking advantage of the SO(2,4) symmetry with which the AdS space is endowed. In this way, the projection π onto the latent manifold, identified by the autoencoder as ∂W, constitutes the inverse of the postulated holographic map $h : \partial W \to W$ (Figure 3.16) that makes the diagram on Figure 3.17 commutative.

Thus, the AdS/CFT equivalence may be proved by AI. The program entails two steps: a) generation of a data-driven holographic metamodel of quantum systems by formulating its supergravity dual/equivalent on a DL NN, and b) construction of the appropriate autoencoder (π, K_{QFT}) that fits the holographic map h, yielding the functional identity $h \circ K_{QFT} \circ \pi = F_{ST}$ of homeomorphisms on W. Part a) of the program has already been achieved, as shown elsewhere [38]. A deep NN representation of the AdS/CFT correspondence has been obtained, with emergence of the bulk metric based on deep learning of data generated as outcome of boundary quantum field theories. The radial direction of the bulk metric is assimilated with the depth of the hidden layers. Thus, the network provides a data-driven model of strongly coupled systems. In the space-time for a black hole horizon, the deep NN can fit boundary data generated by the AdS Schwarzschild space-time, reproducing the metric as the data is reverse engineered. With inputted experimental data, the deep NN determines the bulk metric, mass, and quadratic coupling. Thus, the NN provides a gravitational model of strongly correlated systems.

As for part b), the holographic autoencoder (HAE) that will ultimately identify the latent manifold with ∂W can be obtained using the methods described in this chapter. The remaining challenge in part b) is for the

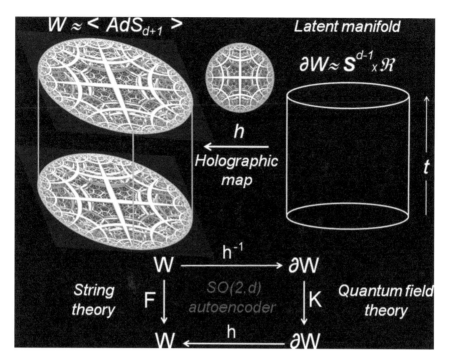

FIGURE 3.16 Holographic autoencoder enabling the discovery of a quantum metamodel of gravity.

$$W \quad \overset{\pi}{\to} \quad \partial W$$

$$\downarrow F_{ST} \qquad \downarrow K_{QFT}$$

$$W \quad \overset{h}{\leftarrow} \quad \partial W$$

FIGURE 3.17 Commutative diagram for a holographic autoencoder.

HAE to simultaneously yield the parsimonious QFT at the boundary ∂W that serves as metamodel for the gravity model on W in the sense adopted in this book. To address this challenge requires that we harness the isomorphism already noted [39] between a deep autoencoder and the multiscale entanglement renormalization *ansatz* (MERA).

If the universe is indeed a hologram, as Maldacena and others suggest [34], the inverse of the holographic map of its avatar, the event horizon of the black hole, represents a 1–1 (injective) projection onto the horizon boundary. Therefore, this boundary may be regarded as a model for the latent quotient manifold of the universe. In this quotient manifold, one dimension has been folded up and stored within the equivalence classes that are ultimately the only observables and hence the only encoded features endowed with physical entity. This topology for the quintessential universe is not entirely correct but locally correct, as shown in Chapter 4. The overall topology of the universe is immaterial to this discussion, as both quantum physics and GR are in fact local theories.

Ultimately, quantum gravity will be ascertained through a holographic autoencoder that identifies the "correct" latent manifold and associated quantum metamodel, or perhaps by other means available to theoretical physicists. Be as it may, we may state that all events following the Planck epoch of the big bang ($t > 10^{-43}$s [40]) are likely to be of a quantum nature. This is so because at that stage, gravity branches off from the three other fundamental forces already accounted for by quantum theory, so all four differentiated forces will be reliably identified as quantum forces. A rigorous topological characterization of the universe supports this assertion, as shown in Chapter 4, with gravity originally stored as a de Broglie wave in a compact fifth dimension. Thus, the certainty of all events that follow the Planck epoch can only be established through the participation of observers. This singular circumstance leaves us having to postulate God's existence at least as early as 10^{-43}s after the big bang, or admitting that the universe remains a mere possibility, replete with phenomena-to-be within multiple *a priori* potential realities that are equally possible in the quantum realm, as in the multiverse scenario.

3.14 AI ENCODES QUANTUM GRAVITY

To address the problem of quantum gravity from an AI–borne perspective, we need to design a holographic autoencoder by leveraging the physics of machine learning to address the problem of whether emergent quantum behavior can arise in a neural network. By emergent quantum mechanics we mean a formulation within a framework of nonlocal hidden variables, as in the Bohm scheme [41]. Once emergent quantum behavior is shown to become possible in machine learning physics, we may address the question of developing a relativistic string gravitational scheme on the hidden variables adopted in a Bohm-inspired quantum network architecture. Thus, the latter becomes in effect a quantum gravity autoencoder for the network materializing emergent gravity.

To develop a network with emergent quantum behavior, we first need to focus on NNs as statistical mechanics entities and closely examine the statistical physics that may underlie generic machine learning. To start, let us consider the $N \times N$ connectivity matrix $\mathbf{w}$ and the bias N-vector $\mathbf{w_0}$ as stochastic variables with entries generically denoted q_i, $i = 1,\ldots, N+N^2$. The NN state vector $\mathbf{x}$ is thus updated in discrete time steps according to the usual scheme of f-activation:

$$x(t+\tau) = f(wx(t)+w_0).$$
(3.26)

Here the time interval τ represents the overall thermalization time for the state vector $\mathbf{x}$, whose entries will be regarded as the hidden variables. To develop the near-equilibrium thermodynamics scheme, let us define a loss function $J(\mathbf{x},\mathbf{q})$ that penalizes departures from the equilibrium which is achieved as $x(t + \tau) \approx x(t)$ for $t \gg \tau$. If μ represents the "reduced mass" of the network, we get

$$J(x,q) = \frac{1}{2}\| x - f(wx+w_0)\|^2 - \mu \| x \|^2 .$$
(3.27)

Thus, the statistical thermodynamics near equilibrium stems directly from the canonical partition function

$$Z(\beta,q) = \int \exp\left[-\beta J(x,q)\right]d^N x.$$
(3.28)

Following the tenets of statistical mechanics, the partition function yields the Helmholtz free energy for the NN given by:

$$A(\beta,q) = -\beta^{-1} \log\left[Z(\beta,q)\right].$$
(3.29)

In the specific case of an activation function given by the hyperbolic tangent, the partition function for the NN may be calculated as:

$$Z(\beta,q) = 2\pi^{N/2} \left\{\det\left[\beta G(w)+(1-\beta\mu)I\right]\right\}^{-1/2}$$
(3.30)

where $G(w)$ is given by [42]:

$$G(w) = (I - f'w)^T (I - f'w),$$
(3.31)

where f' is the diagonal matrix of first derivatives of the activation function with respect to each component of the NN state vector x.

When the network is at thermodynamic equilibrium, the average loss $\langle J(x, q) \rangle = U(\beta, q)$ becomes a minimum, hence the Helmholtz free energy becomes

$$A(\beta,q) = \frac{1}{2}\beta^{-1}\left[-N\log(2\pi) + \sum_{\lambda_i > \beta^{-1}}\log(\lambda_i)\right] + \tilde{A}(\beta) \qquad (3.32)$$

where λ_i are the eigenvalues of $G(w)$ and $\tilde{A}(\beta) = U(\beta) - \beta^{-1}S(\beta)$ is the thermodynamic Helmholtz free energy of the NN.

To describe the thermodynamic behavior of the NN near equilibrium, we take into account that the entropy production is stationary and introduce the time-dependent probability distribution $p(t, q)$ with Shannon entropy $S(t, q) = -\int p(t, q) \log [p(t, q)]dq$. Assuming the learning evolutionary drift follows (i.e., is proportional to) the gradient of the free energy, we get $\dfrac{dq_j}{dt} = \dfrac{\varsigma \partial A}{\partial q_j}$, hence the minimum entropy production yields the set of constraints

$$\frac{\partial A}{\partial t} + \varsigma\left(\frac{\partial A}{\partial q_j}\right)^2 = \left\langle\frac{dA(t,q)}{dt}\right\rangle_t. \qquad (3.33)$$

This set of equations in turn begets a minimal action principle defined variationally as

$$\frac{\delta J(p,A)}{\delta p} = \frac{\delta J(p,A)}{\delta A} = 0, \qquad (3.34)$$

where the action $J(q,A)$ becomes

$$J(p,A) = \int_0^\infty \frac{dS(t,q)}{dt}dt$$

$$+ \vartheta \int\int_{t=0}^{\infty} p(t,q)\left\{\frac{\partial A}{\partial t} + \sum_{j=1}^{N+N^2}\left[\varsigma\left(\frac{\partial A}{\partial q_j}\right)^2\right] - \left\langle\frac{dA(t,q)}{dt}\right\rangle_t\right\}dtdq \qquad (3.35)$$

with ϑ denoting the corresponding Lagrange multiplier.

The action may be rewritten taking into account the Fokker-Planck equation satisfied by $p(t, \boldsymbol{q})$:

$$\frac{\partial p}{\partial t} = \frac{\partial}{\partial q_j}\left[\frac{D\partial p}{\partial q_j} - \frac{\zeta\partial A}{\partial q_j}p\right] \tag{3.36}$$

where the parameter D plays the role of diffusion coefficient for the NN dynamics near equilibrium. Taking into account Equation (3.36) in the computation of the time derivative of the Shannon entropy $S(t, \boldsymbol{q})$ enables us to rewrite Equation (3.35), so the action now reads:

$$\mathcal{J}(p, A) = \int\limits_{t=0}^{\infty}\int p(t,\boldsymbol{q})\left\{\vartheta\frac{\partial A}{\partial t} + \sum_{j=1}^{N+N^2}\left[\zeta\frac{\partial^2 A}{\partial q_j^2} - 4D\frac{\partial^2 p}{\partial q_j^2} + \vartheta\zeta\left(\frac{\partial A}{\partial q_j}\right)^2\right]\right.$$
$$\left. -\vartheta\left\langle\frac{dA(t,\boldsymbol{q})}{dt}\right\rangle_t\right\}dtd\boldsymbol{q}. \tag{3.37}$$

This action can be written equivalently in the form of a Schrödinger action:

$$\mathcal{J}(p, A) = \int\limits_{t=0}^{\infty}\int \Psi^*\left[-4D\sum_{j=1}^{N+N^2}\frac{\partial^2}{\partial q_j^2} - i\eta\frac{\partial}{\partial t} + \Upsilon(\boldsymbol{q})\right]\Psi dtd\boldsymbol{q} \tag{3.38}$$

with $\eta = \left(\frac{4D}{\zeta}\right)^{1/2}$, $\Upsilon(\boldsymbol{q}) = -\left\langle\frac{dA(t,\boldsymbol{q})}{dt}\right\rangle$ and wave function

$$\Psi = p^{1/2}\exp\left[i\eta^{-1}A\right]. \tag{3.39}$$

Thus, naming $\eta = \hbar$, we obtain the Schrödinger equation describing the state of the system as a particle-wave in the q-space $\mathbb{R}^{N\times N} \times \mathbb{R}^N$ of trainable variables for the learning process near equilibrium with state x-vector entries represented as thermalized hidden variables:

$$i\eta\frac{\partial}{\partial t}\Psi = \left[-4D\nabla^2 + \Upsilon(\boldsymbol{q})\right]\Psi. \tag{3.40}$$

We have surveyed the statistical thermodynamics of an NN with stochastic trainable variables. The system evolves with the appropriate time coarse graining associated with the thermalization limit for hidden Bohm variables representing entries in the state vector. Such a system is capable of exhibiting an emergent quantum behavior. We are now ready to address the crucial question that underlies the quantum gravity conundrum from the standpoint of AI:

Can this quantum system be regarded as the variational autoencoder of an NN with emergent gravity in a Minkowski space-time?

To address this question, we need to consider the projection $\pi_\tau: x \rightarrow \bar{x}(q)$, where $\bar{x}(q)$ is the equilibrated state vector of the NN relative to the specific realization q of the stochastic trainable variables that in turn evolve in multiples of the equilibration time t. In rigorous terms we get:

$$\bar{x}(q) = \int x \exp\left[-\beta J(x,q)\right] d^N x. \tag{3.41}$$

Thus, we are enquiring whether it is possible to treat the nonequilibrium hidden variables (entries) in state vector x by mapping relativistic strings in an NN with an emergent Minkowski space, so that the entropy production in such a system is a function of the metric tensor that describes weak interactions between training subsystems of x-values. The answer is affirmative because Equations (3.30) and (3.31) may be specialized to the case where the weight vector $\mathbf{w}$ is simply a permutation matrix Ξ with an arbitrary number of cycles [42, 43], so that the matrix $\mathbf{G}$ now becomes

$$\mathbf{G} = (\Xi - \mathbf{I})^T (\Xi - \mathbf{I}). \tag{3.42}$$

Thus, the stochastic NN with partition function

$$Z(\beta,\Xi) = 2\pi^{N/2} \left\{\det\left[\beta(\Xi - \mathbf{I})^T (\Xi - \mathbf{I}) + (1 - \beta\mu)\mathbf{I}\right]\right\}^{-1/2} \tag{3.43}$$

represents a quantum gravity autoencoder for the NN with emergent relativistic gravity, so that the diagram in Figure (3.18) becomes commutative.

The emergent quantum behavior was shown to average out or thermalize the hidden variables that have been identified as components of the state x-vector of the network. Thus, the trainable variables conforming the q-vector were shown to exhibit a quantum mechanical behavior in an equilibrium regime. We assume without loss of generality that the learning process involves L separate sets of training $\mathbf{x}$-vectors with expected values $\bar{x}^l, l = 1, 2, \ldots, L$ and these expectation vectors, together with the overall expectation vector ($l = 0$) representing the sum of all L expectation training vectors, are regarded as the hidden variables in the emergent quantum behavior of the trainable $\mathbf{q}$-states. On the other hand, the nonequilibrium dynamics of the hidden variables becomes relevant on timescales much smaller than their thermalization time. This nonequilibrium dynamics is determined by the strength of the weak interactions between vector pairs $\bar{x}^\nu, \bar{x}^\xi, \nu, \xi = 0, 1, \ldots, L$ quantified by the tensor $g^i_{\nu\xi}$, where the dummy index i labels each neuron in the system.

To endow the hidden variables with an emergent gravity action, we first describe the nonequilibrium dynamics of the expectation vectors for the training sets. For the sake of transparency, we assume the simplest possible activation function $f = I$. Thus, for $\Delta t = \tau_< \ll \tau$, we get to first order:

$$\bar{x}^\mu_i(t + \tau_<) \approx w_{ij}\bar{x}^\mu_j(t), \mu = 0, 1, \ldots, L. \tag{3.44}$$

This nonequilibrium scheme yields a tangent bundle according to

$$\frac{\partial \bar{x}^\mu_i}{\partial t} \approx \tau_<^{-1}\left[w_{ij} - \delta_{ij}\right]\bar{x}^\mu_j(t). \tag{3.45}$$

And since $\bar{x}^0 = \sum\limits_{l=1}^{L}\bar{x}^l$, we get

$$g^i_{\mu\nu}\frac{\partial \bar{x}^\mu_i}{\partial t}\frac{\partial \bar{x}^\nu_i}{\partial t} = 0 \tag{3.46}$$

where the metric tensor $g^i_{\mu\nu}$ describes the magnitude of the interactions between the hidden variables now cast in terms of the differential geometry of the emergent space-time.

The interactions between different training sets arise from the loss function $J_<(\mathbf{q})$ that holds for timescales shorter than the equilibration time for the hidden variables. This loss function becomes:

$$J_<(\mathbf{q}) = J_<(\mathbf{w}) = \sum\limits_{l=0}^{L}J^{(l)}_<(\mathbf{w}) = \sum\limits_{l=0}^{L}\sum\limits_{x^l \in S_l}\sum\limits_{n=1}^{M=\left[\frac{\tau}{2\tau_<}\right]}\left\|x^l\left((n+1)\tau_<\right) - wx^l\left(n\tau_<\right)\right\|^2 \tag{3.47}$$

where $S^l\left(l=0,\ldots,L\right)$ is the l-th training set. Thus, we define the metric tensor as

$$g^i_{\mu\nu} = \left[\mathrm{Argmin}\left(J^{(\mu)}_<\right) - \mathrm{Argmin}\left(J^{(\nu)}_<\right)\right]^{-1}_i \tau^{-1}\sqrt{\int_{t=0}^{\tau}\left\|\bar{x}^\mu(t) - \bar{x}^\nu(t)\right\|^2 dt}$$

(3.48)

The gravity action then becomes:

$$J\left(w\right) = g^i_{\mu\nu}\left[\tau_<^{-2}\langle x^\mu_i G_{ij} x^\nu_j\rangle - \frac{\partial\bar{x}^\mu_i}{\partial t}\frac{\partial\bar{x}^\nu_i}{\partial t}\right]$$

(3.49)

with $G = G(w) = (I - w)^T(I - w)$.

Or in Einstein's relativity terms:

$$J\left(w\right) = \int dX\sqrt{-g}\,g_{\mu\nu}T^{\mu\nu},$$

(3.50)

where $g = \det\left(g_{\mu\nu}\right)$ and

$$\sqrt{-g}T^{\mu\nu} = \left[\tau_<^{-2}\langle x^\mu_i G_{ij} x^\nu_j\rangle - \frac{\partial\bar{x}^\mu_i}{\partial t}\frac{\partial\bar{x}^\nu_i}{\partial t}\right]\prod_{l=0}^{L}\delta\left(X^l_i - \bar{x}^l_i\right).$$

(3.51)

Equations (3.47)–(3.51) define the emergent gravity of the neural network arising from the nonequilibrium dynamics of the hidden variables in the quantum mechanical autoencoder.

3.15 RELATIVISTIC STRINGS IN A PHYSICAL REALIZATION OF THE QUANTUM GRAVITY AUTOENCODER

The statistical thermodynamics of machine learning is currently being elucidated by turning nontrainable (x) and trainable (q) variables into the stochastic variables for the NN and its variational autoencoder, respectively [42]. As demonstrated previously, an NN may be endowed with emergent gravity while its autoencoder is governed by a latent Schrödinger Equation (3.40), thus exhibiting a quantum behavior. To generate this metamodel of quantum gravity, it is necessary to a) treat the nontrainable variables as hidden variables in the emerging quantum gravity autoencoder, b) consider a limit where the weight matrix (*w*) becomes a permutation matrix, and c) treat the hidden variables in a nonequilibrium setting on timescales shorter than thermalization times by generating subsystems of state vectors whose dynamics are described by relativistic strings in an emergent Minkowski space-time. The latent manifold associated with the Minkowski space-time is then obtained by thermalization of the hidden variables. The relativistic strings become enslaved or entrained in the thermalization limit where the nontrainable variables are treated as equilibrated vis-à-vis the trainable variables and as such they are subsumed in the latent Schrödinger equation via the Helmholtz free energy.

Thus, we may conclude by stating that AI provides a quantum metamodel of gravity, and hence the big bang is in all likelihood a quantum event. In this context at least one of the following four statements is valid:

A) The laws of quantum mechanics cannot be upheld in the big bang setting.

B) A quantum tunneling event generated the universe as progeny of another universe. The tunneling occurred across the barrier separating two quantum gravity autoencoders with respective latent manifolds Ω and Ω^*. The amplification of the information tunneled to the latent manifold Ω^* was realized as the funneled decoding of the latent wavefunction, giving rise to the progeny universe (Figure 3.19).

C) The *a priori* presence of a primeval observer "Sensus Dei" materialized the big bang, which therefore is not a phenomenon-to-be in Wheeler's sense [44] but a realized event.

D) The big bang is a phenomenon-to-be in Wheeler's sense and hence we are part of a multiverse, with the universe as possibility.

3.16 THE UNIVERSE AS A HOLOGRAPHIC AUTOENCODER

This chapter addressed the conundrum of quantum gravity by reductively regarding the universe as the realization of a learning system with stochastic weights and biases where gravity and quantum behavior become emergent properties within the physics of machine learning. To that effect, we explore the possibility of an AI–based construction of a quantum holographic autoencoder which requires that the emergent quantum behavior arises in a neural network. Once an emergent quantum behavior is shown to become possible within the machine learning system equilibrated on the nontrainable hidden variables, we address the question of developing a relativistic string gravitational scheme on the hidden variables adopted. Thus, the network with equilibrated nontrainable variables becomes in effect a quantum gravity autoencoder for the underlying network exhibiting emergent gravity in the nonequilibrium regime prior to the equilibration of the nontrainable variables. In this way we build a quantum metamodel for gravity that fulfills at least in part a major imperative for physicists seeking a unified field theory. Furthermore, the physical possibility of tunneling across quantum gravity autoencoders supports the idea that our universe may be the progeny of an older universe [45] that dreamt – or simulated – it.

This is a bold claim, yet it may be deconstructed vis-à-vis the main objective of this chapter, which was to describe the behavior of the neural networks in the limit where the bias vector, weight matrix, and state vector of neurons can be modeled as stochastic variables that undergo a learning evolution. As it turns out, this learning evolution, when projected onto the autoencoder, can be described by the time-dependent Schrödinger Equation (5.40), and the time evolution dictated by this equation is compatible with the relativistic decoding enshrined in the commutativity of the diagram presented in Figures 3.17 and 3.18. Taken together, these results have clear implications for the possible emergence of quantum mechanics, general relativity, and mesoscopic observers in neural networks governed by a unified theoretical scheme that adopts two different guises in the two different thermodynamic regimes (Figure 3.18).

$$x(t) \quad \overset{\pi_\tau}{\to} \quad \overline{x}(t, \Xi)$$

$$\downarrow F_{ST} \qquad \downarrow K_{QM}$$

$$x(t + \tau) \quad \overset{\pi_\tau}{\to} \quad \overline{x}(t + \tau, \Xi)$$

FIGURE 3.18 Commutative diagram for a quantum gravity autoencoder, with F_{ST}, K_{QM} representing respectively the string and quantum flow map.

Thus, our construct upholds the controversial view that quantum mechanics may not be a fundamental theory, but rather an *ansatz* giving rise to a mathematical tool that allows us to carry out statistical calculations with great efficacy and accuracy in a certain class of dynamical systems. In this guise, an emergent quantum mechanics should be derivable from the first principles of statistical mechanics. This is precisely what this chapter has accomplished for a dynamical system consisting of a neural network that is in effect a learning system that contains two different types of degrees of freedom: The trainable bias vector and weight matrix elements and the nontrainable state vector of neurons, with the latter constituting the hidden variables.

Emergent gravity is also a relatively new area of research, but in this case the picture is far more nebulous than in emergent quantum mechanics: It is far less clear whether progress has been made, if at all. The main hurdle is that emergent gravity requires also an emergent space, an emergent Lorentz invariance, and an emergent general relativity [46, 47]. To our surprise, the string-theory–based nonequilibrium treatment of the hidden variables in neural networks opened up a window of opportunity to treat the conundrum of emergent gravity in a completely unified fashion that encompasses all three aspects of the aforementioned problem in context of the learning dynamics. As it appears to be the case, a relativistic space-time can indeed emerge from a nonequilibrium evolution of the hidden variables in a manner that is very much akin to string theory [42]. More specifically, as described by Vanchurin [42], if one considers D minimally interacting subsystems (through bias vector and weight matrix) with average state vectors, then the emergent dynamics can be modeled with relativistic strings in an emergent D + 1 dimensional Minkowski space-time. Furthermore, the emergent dynamics may be modeled with the Einstein equations provided the weak subsystem interactions are described by a metric tensor. In this way, a stochastic learning dynamics scheme such as the one proposed in this chapter proved to be instrumental for the equilibration of the emergent space-time that turned out to exhibit a behavior describable by a gravitational theory such as general relativity.

3.17 COSMIC REPRODUCTION BY QUANTUM TUNNELING ACROSS COUPLED QUANTUM GRAVITY AUTOENCODERS

The previous discussion addresses one of the biggest problems concerning the history of the universe: What happened before the big bang? One may recall that Albert Einstein was never satisfied with the big bang scenario itself because he thought that a beginning in time would need to be postulated in a seemingly *ad hoc* manner. This way of doing physics to him was unacceptable, symptomatic of a feeble theory. Additionally, his own space-time would make a "beginning" conceptually inconceivable as the clock is subsumed in the manifold itself, rendering the evolutionary picture meaningless: If the universe is space-time, the universe evolution is contained in the universe itself, yielding a Russell-type paradox.

After nearly a century since Einstein voiced his skepticism – which extended to the whole of quantum mechanics as well – a whole gamut of hypotheses have been formulated regarding our cosmic origin. Perhaps the most sound include ideas such as the following: a) The universe sprout from a quantum vacuum fluctuation, b) the universe involves infinite cycles of contraction and expansion, c) the universe was selected through the anthropic principle stemming from the string theory landscape of the multiverse, where every possible event or phenomena is implicitly encompassed, none materialized, and the big bang itself is a phenomenon-to-be in Wheeler's sense, and d) the universe emerged from the collapse of matter in the interior of a black hole that was contained in a progenitor universe.

Be as it may, none of these ideas can be ascribed full credibility, mainly because none of the theories they stem from has satisfactorily solved the conundrum of quantum gravity. A less explored and more daring possibility put forth in this chapter is that *our universe was created in the laboratory by a technologically advanced civilization capable of harnessing the power of quantum gravity autoencoders*. This requires a mastery of the physics of learning machines and an ability to craft gravity and quantum behavior as emergent attributes of a stochastic learning system that admits a quantum gravity autoencoder (Figure 3.17). The underlying theory behind this idea does not portend to solve the quantum gravity conundrum *per se* but at least reconciles the two main forces as emergent in a single learning machine. Furthermore, since our universe is endowed with a flat geometry at a zero net energy (Chapter 4), an advanced civilization could have harnessed the power of quantum gravity autoencoders to create a baby universe through quantum tunneling [45] into a second quantum gravity autoencoder acting as antenna and reservoir for the spillover probability, as schematically depicted in Figure 3.19.

This "emergent matrix" idea of the origin of the universe reconciles the theological need for a "creator," that is, the primeval quantum observer

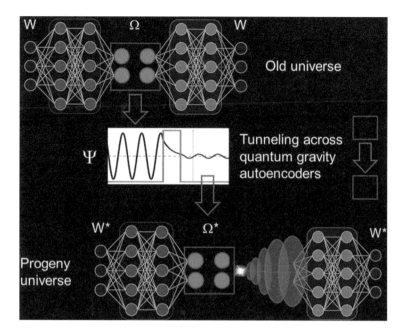

FIGURE 3.19 Schematic representation of a tunneling event across two quantum gravity autoencoders generating the universe (W*, Ω*) as progeny of an older universe (W, Ω). The amplification of information tunneled to the latent manifold Ω* in a quantum spillover event materializes as the relativistic decoding of the latent wavefunction, an act of creation event giving rise to the progeny universe.

that would have bestowed reality to the big bang by detecting the event, with the "a priori" secular concept of quantum gravity. As said, while we have not gotten a cogent theory that conceptually unifies quantum mechanics and gravity, we have a "matrix framework of the universe," a neural network architecture where the two key forces in modern physics are reconciled, so that quantum behavior in the autoencoder can be decoded back as gravity through the holographic map that constitutes the inverse of the canonical projection. Surely a more advanced civilization that mastered the technology of quantum gravity autoencoders would be able to accomplish the feat of creating baby universes leveraging physical autoencoders or other equivalent vehicles for quantum tunneling. If so, our universe was not selected for us to dwell in it and bestow reality to the quantum events we are capable of detecting – as upheld by the standard anthropic principle – but rather, it was selected to host civilizations that are much more technologically advanced than we are. These civilizations capable of leveraging the technology required to create progeny universes, be it quantum gravity autoencoders or some unfathomable alternative vehicle efficacious at harnessing quantum tunneling, would be the actual drivers of the cosmic selection process.

By contrast, we are incapable at this time of harnessing technology for cosmological manipulation, and obviously we are incapable of recreating the cosmic conditions that led to our existence. In plain words, our civilization is cosmologically still at a rudimentary stage since we do not possess the technology to reproduce the universe that has hosted us for quite a while already. If we were to measure the technological level of a civilization by the ability to recreate or reproduce the astrophysical conditions that led to its existence, we would say that we are at early stages of development, possibly a low-level civilization, graded class C on a cosmic scale. By contrast, a civilization in the A class rank could recreate the astrophysical conditions that gave rise to its existence, namely produce a baby universe through a quantum-controlled laboratory experiment that leverages a tunneling effect through an appropriately crafted vehicle. A class A civilization would also be able to effectively address related challenges, such as producing a large enough density of dark energy to hold the universe together after its inception, as has already been discussed in the scientific literature [45].

However sound and cogent, the theories on the quantum origins of the universe proposed so far [48] can be subject to a common and basic criticism: *There is no certainty that the universe can be treated as a quantum object.* This chapter heralds an improvement of this state of affairs as the duality enshrined in quantum gravity is realized through a purposely built autoencoder that compresses gravitational string data as emergent quanta.

In accord with Equation (3.40), if our universe is to become information-compressed within the quantum gravity autoencoder, the following relations must hold: $\zeta = \dfrac{1}{2m}, D = \dfrac{\hbar^2}{8m}$, or reciprocally:

$$\hbar = 2\sqrt{\frac{D}{\zeta}}, \tag{3.52}$$

begetting the uncertainty relation adapted for the quantum gravity autoencoder:

$$\Delta E \times \Delta t \sim \sqrt{\frac{D}{\zeta}}. \tag{3.53}$$

While the total energy of the NN with emergent gravity is zero (a closed universe has zero energy [49]), the total energy of its quantum gravity autoencoder is $U = -\dfrac{\partial}{\partial \beta} \log Z(\beta, q) \neq 0$ in accord with the tenets of statistical mechanics. This paradox may be resolved by noting that the observational timescales for the gravitational NN and its autoencoder are different. Thus, in accord with the uncertainty principle, the total energy of the autoencoder may be zero for a timespan Δt which is incommensurably shorter than the equilibration time τ for the hidden variables. This implies that the parameter τ needs to be tuned as an architectural determinant of the autoencoder, so that the following relation is fulfilled:

$$\tau \gg \Delta t \sim \sqrt{\frac{D}{\varsigma}} \left| \frac{\partial}{\partial \beta} \log Z(\beta, q) \right|^{-1}. \tag{3.54}$$

Thus, the incommensurability of the equilibration timescale relative to the timescales associated with the hidden variables modelled with relativistic strings implies that the universe may be a vacuum quantum fluctuation. This possibility is allowed by the uncertainty principle as described by Equation (3.54).

We should emphasize that the possibility that the universe as intelligible information is actually a vacuum quantum fluctuation is not as far-fetched as it may seem. A simple back-of-the-envelope calculation involving the cosmological constants of our known universe leads to an equivalent result. Thus, the energy $E = mc^2$ of a material object of mass m is actually counterbalanced by its gravitational potential energy $E_g = -GmM/R$, where G is the gravitational constant and M is the net mass of the universe contained within the Hubble ball of radius $R = c/H_0$, where H_0 is the Hubble constant [49]. To prove the previous assertion, we note that the critical minimal mass contained in the Hubble ball of volume $\dfrac{4}{3}\pi \left(\dfrac{c}{H_0}\right)^3$ and required for the universe to be closed is $M = c^3/(2GH_0)$, implying that the gravitational energy E_g compensates the energy E up to a constant of order unity.

The results described in this section pave the way for a cosmological technology that harnesses AI, or more specifically, the power of quantum gravity autoencoders. Thus, two parameters, D and τ, may be tuned to harness the power of AI to manipulate cosmological scales to the point of giving birth to a universe that serves as a metamodel for emergent quantum gravity.

The technological feat of cosmic reproduction implemented as shown in Figure 3.19 would not be completely described without a proper space-time rendering of the act of creation. Thus, the quantum tunneling event triggering cosmic reproduction should be represented in a cross section of the space-time manifold of the progenitor universe, as schematically delineated – with limitations – in Figure 3.20. As shown in Chapter 5, the latent manifold of our universe is – at least intermittently – a 4-torus with unity aspect ratios (4-dimensional horn torus). Hence, given the connection at the point of extreme curvature, the latent universe contains a primeval wormhole, a bridge connecting the interiors of two entangled black holes (BHs) (Figure 3.21). Quantum tunneling generates an entanglement (Chapter 5) of the connected BHs of the progenitor universe with the primeval BHs of the progeny universe. In the particular space-time cross section on display in Figure 3.20, only one of the primeval BHs from each universe is shown to be connected through an "inter-universe" Einstein-Rosen bridge. However, in

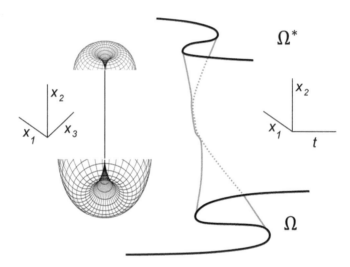

FIGURE 3.20 The quantum tunneling for cosmic reproduction represented in a cross section of the space-time manifold. As shown in Chapter 5, a toroidal universe with unity aspect ratios (4-dimensional horn torus) contains a primeval wormhole, a bridge connecting the interiors of two entangled black holes (BHs). The quantum tunneling generates an entanglement (Chapter 5) of the connected BHs with the primeval BHs of the progeny universe. This entanglement materializes as an Einstein-Rosen bridge or wormhole connecting the respective primeval wormholes of parental and progeny universe. In the particular space-time cross section on display, only one of the primeval BHs from each universe is shown to be bridged.

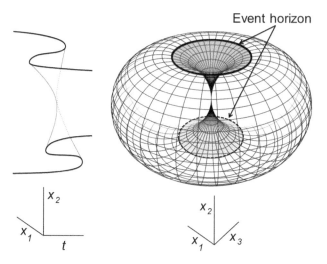

FIGURE 3.21 The space-time rendering of the primeval wormhole in a 4-torus universe with unitary aspect ratios.

a more accurate depiction, the entanglement arising from the quantum tunneling should actually materialize as a "supra-wormhole" connecting the primeval wormholes of parent and progeny universe, as depicted in Figure 3.22.

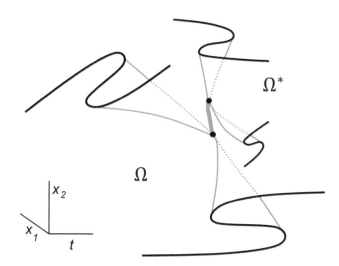

FIGURE 3.22 Space-time rendering of the quantum entanglement of the two primeval wormholes associated with the creation of a progeny universe through quantum tunneling. The latent space-times for the parental and progeny universe are indicated respectively as Ω, Ω^*.

The act of creation schematically depicted in Figure 3.22 is undoubtedly perplexing and attests to the possibility of superior technology capable of endowing an AI system, that is, a holographic autoencoder, with physical entity, as described in this chapter. This act of creation would materialize by coupling two such autoencoders, so one of them serves as the antenna for quantum tunneling when interfacing with the one enacting the progenitor universe.

There is *a priori* no way to tell whether we live/exist in the parental universe or in a parentally entangled progeny universe generated by a superior civilization that exists or has existed in the parental universe and is or was capable of realizing the feat described in Figure 3.22. The conundrum is akin to Jorge Luis Borges' fiction "The Circular Ruins," where a man meticulously sets himself up to dream another man, only to realize that he himself is actually being dreamt by someone else.

3.18 COSMIC REPRODUCTION AS A PHYSICAL IMPOSSIBILITY

The previous discussion paved the way to implement a birth channel for a progeny universe by amplifying a latent quantum fluctuation originated in a quantum gravity autoencoder and received by another autoencoder acting as antenna. This fluctuation is transmitted as a quantum tunneling event into the purported progeny autoencoder of a wave function that spans the parent-universe autoencoder. The physicality of such an event is seemingly impossible as we shall presently show, unless we assume, as done in Chapter 6, that our universe is the outcome of a computer simulation. We must seek for the culprit of the physical impossibility in the topology of space-time as we currently understand it (Chapter 2).

The inherently local nature of general relativity did not prompt Einstein to make any specific assumption on the topology of space-time. Since the inception of Einstein's theory of gravity, it was usually assumed that any spatial cross section of space-time is a 3-dimensional ball $\left(\mathcal{S}^3 \subset \mathbb{R}^3\right)$. As such, the boundary of the universe would represent a barrier bestowing meaning to the idea that a universe may sprout from the tunneling of a quantum fluctuation originated in a parental universe [50]. However, we have asserted on physical grounds that the spatial cross sections of space-time are best represented as compact multiply connected manifolds homeomorphic to the 3-dimensional torus $T^3 \approx \mathbb{R}^3/\mathbb{Z}^3$. That is, *there is no boundary, and hence no barrier to the parent universe*, as depicted in Figure 3.23. A barrier would necessarily imply that there is geometry

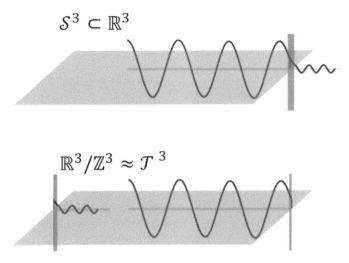

$$S^3 \subset \mathbb{R}^3$$

$$\mathbb{R}^3 / \mathbb{Z}^3 \approx \mathcal{T}^3$$

FIGURE 3.23 Physical impossibility of the spillover of a quantum fluctuation originated in a parental universe assuming a compact multiply connected topology for spatial cross sections of space-time. In that case, accepted as accurate depiction of the universe topology, the probability distribution of each and every quantum fluctuation is completely contained in the parental universe, where it originated.

beyond it, that is, as we cross the barrier, but in the toroidal universe (or its non-orientable version described in Chapter 5), there is no geometry outside space-time: There is nothingness. A quantum fluctuation cannot spill over from the parent universe to yield a progeny universe because there is no barrier/boundary to bestow entity to the tunneling event: Each and every quantum fluctuation would have its probability distribution 100% contained in the parent universe.

REFERENCES

1. Russell S, Norvig P (2020) *Artificial Intelligence: A Modern Approach.* Pearson, London, UK.
2. Kelleher JD (2019) *Deep Learning.* The MIT Press, Cambridge, MA.
3. Chollet F (2019) *Deep Learning with Python.* Manning Press, Shelter Island, NY.
4. Atienza R (2020) *Advanced Deep Learning with TensorFlow 2 and Keras*, 2nd Edition. Packt Publishing, Birmingham, UK.
5. Schmidhuber (2015) Deep learning in neural networks: An overview. *Neural Networks* 61: 85–117.
6. Fernández A (2016) *Physics at the Biomolecular Interface.* Springer International Publishing, Switzerland.

7. Aliper A, Plis S, Artemov A, Ulloa A, Mamoshina P, Zhavoronkov A (2016) Deep Learning Applications for Predicting Pharmacological Properties of Drugs and Drug Repurposing Using Transcriptomic Data. *Mol Pharm (ACS)* 13: 2524–2530.

8. Lavecchia A (2019) Deep Learning in Drug Discovery: Opportunities, Challenges and Future Prospects. *Drug Disc Today* 24: 2017–2032.

9. Jiménez J, Škalič M, Martínez-Rosell G, De Fabritiis, G (2018) Kdeep: Protein–Ligand Absolute Binding Affinity Prediction via 3D-Convolutional Neural Networks. *J Chem Inf Mod* 58: 287–296.

10. Stepniewska-Dziubinska MM, Zielenkiewicz P, Siedlecki P (2018) Development and Evaluation of a Deep Learning Model for Protein–Ligand Binding Affinity Prediction. *Bioinformatics* 34: 3666–3674.

11. Hassan-Harrirou H, Zhang C, Lemmin T (2020) RosENet: Improving Binding Affinity Prediction by Leveraging Molecular Mechanics Energies with an Ensemble of 3D Convolutional Neural Networks. *J Chem Inf Model* 60: 2791–2802.

12. Brunton SL, Kutz NJ (2019) *Data-Driven Science and Engineering: Machine Learning, Dynamical Systems and Control.* Cambridge University Press, UK.

13. Tiumentsev Y, Egorchev M (2019) *Neural Network Modeling and Identification of Dynamical Systems.* Academic Press, San Diego, USA.

14. Champion K, Lusch B, Kutz JN, Brunton SL (2019) Data-Driven Discovery of Coordinates and Governing Equations. *Proc Natl Acad Sci USA* 116: 22445–22451.

15. Schmidt M, Lipson H (2009) Distilling Free-Form Natural Laws from Experimental Data. *Science* 324: 81–85.

16. Fernández A (1985) Center-Manifold Extension of the Adiabatic-Elimination Method. *Phys Rev A* 32: 3070–3076.

17. Born M, Oppenheimer JR (1927) Zur Quantentheorie der Molekeln. *Ann Physik* 389: 457–484.

18. Fernández A (2014) Chemical Functionality of Interfacial Water Enveloping Nanoscale Structural Defects in Proteins. *J Chem Phys* 140: 221102.

19. Fernández A (2021) *Artificial Intelligence Platform for Molecular Targeted Therapy: A Translational Approach.* Chapter 9. World Scientific Publishing, Singapore.

20. Fernández A (2020) Deep Learning Unravels a Dynamic Hierarchy While Empowering Molecular Dynamics Simulations. *Ann Physik (Berlin)* 532: 1900526.

21. Brunton SL, Proctor JL, Kutz NJ (2016) Discovering Governing Equations from Data by Sparse Identification of Nonlinear Dynamical Systems. *Proc Natl Acad Sci USA* 113: 3932–3937.

22. Kurzweil R (2006) *The Singularity Is Near: When Humans Transcend Biology.* Penguin, New York, NY.

23. de Vries J (2014) *Topological Dynamical Systems: An Introduction to the Dynamics of Continuous Mappings.* De Gruyter, Berlin.

24. Duraisamy K, Iaccarino G, Xiao H (2018) Turbulence Modeling in the Age of Data. *Annu Rev Fluid Mech* 51: 357–377.

25. Thommen M, Holtkamp W, Rodnina MV (2017) Co-Translational Protein Folding: Progress and Methods. *Curr Opin Struct Biol* 42: 83–89.

26. Sorokina I, Mushegian A (2018) Modeling Protein Folding In Vivo. *Biology Direct* 13: 13.

27. Thurston WP (1997) *Three-Dimensional Geometry and Topology*. Princeton University Press, Princeton, NJ.

28. Brooks B, Karplus M (1983) Harmonic Dynamics of Proteins: Normal Modes and Fluctuations in Bovine Pancreatic Trypsin Inhibitor. *Proc Natl Acad Sci U S A* 80: 6571–6575.

29. Thirumalai D, Lorimer GH, Hyeon C (2020) Iterative Annealing Mechanism Explains the Functions of the GroEL and RNA Chaperones. *Protein Sci* 29: 360–377.

30. Nemytskii VV, Stepanov V (2016) *Qualitative Theory of Differential Equations*. Princeton University Press, Princeton, NJ.

31. Arnold VI (1974) *Mathematical Methods of Classical Mechanics*. Springer, Berlin.

32. Weisstein EW (2021) Quotient Space. From MathWorld – A Wolfram Web Resource. https://mathworld.wolfram.com/QuotientSpace.html

33. Li S, Yang Y (2021) Hierarchical Deep Learning for Data-Driven Identification of Reduced-Order Models of Nonlinear Dynamical Systems. *Nonlinear Dyn* 105: 3409–3422.

34. Maldacena J (1999) The Large N Limit of Superconformal Field Theories and Supergravity. *Int J Theor Phys* 38: 1113.

35. Witten E (1998) Anti de Sitter Space and Holography. *Adv Theor Math Phys* 2: 263–291.

36. Bekenstein JD (1973) Black Holes and Entropy. *Phys Rev D* 7: 2333–2346.

37. Fernández A, Sinanoglu O (1982) The Lifting of an Inonu-Wigner Contraction at the Level of Universal Coverings. *J Math Phys* 23: 2234.

38. Hashimoto K, Sugishita S, Tanaka A, Tomiya A (2018) Deep Learning and the AdS/CFT Correspondence. *Phys Rev D* 98: 046019.

39. Matsueda H, Ishihara M, Hashizume Y (2013) Tensor Network and a Black Hole. *Phys Rev D* 87: 066002.

40. Weinberg S (2008) *Cosmology*. Oxford, UK: Oxford University Press.

41. Bohm D (1962) A Suggested Interpretation of the Quantum Theory in Terms of Hidden Variables I. *Phys Rev* 86: 166–179.

42. Vanchurin V (2021) Toward a Theory of Machine Learning. *Mach Learn Sci Technol* 2: 036012.

43. Gubser SS (2010) *The Little Book of String Theory*. Princeton, NJ: Princeton University Press.

44. Wheeler JA, Zurek WH (2014) *Quantum Theory and Measurement*. Princeton, NJ: Princeton University Press.

45. Farhi E, Guth A, Guven J (1990) Is It Possible to Create a Universe in the Laboratory by Quantum Tunneling? *Nuc Phys B* 339: 417–490.

46. Vanchurin V (2018) Covariant Information Theory and Emergent Gravity. *Int J Mod Phys A* 33: 1845019.

47. Bednik G, Pujolas O, Sibiryakov S (2013) Emergent Lorentz Invariance from Strong Dynamics: Holographic Examples. *J High Energy Phys* 11: 64.

48. Loeb A (2021) Was Our Universe Created in a Laboratory? *Scientific American*. October 15, 2021. https://www.scientificamerican.com/article/was-our-universe-created-in-a-laboratory/

49. Harrison ER (2003). *Masks of the Universe*. Cambridge University Press, UK.

50. Loeb A (2021) Endless Creation Out of Nothing. *Scientific American* 4: December 12, 2020.

CHAPTER 4

Dark Matter in a Quintessential Extension of the Standard Model

"To be is to be encoded".
ARIEL FERNÁNDEZ

Cosmological data from observation at very large scales pose a significant challenge to the fundamental tenets of physics. Evidence supporting the existence of dark matter (DM) and dark energy (DE) has proven so far unyielding to any possibility of reconciliation with quantum mechanics or relativity. This chapter addresses the problem by resorting to artificial intelligence (AI) for answers and describes the outcome of this investigation in terms of an ur-universe with toroidal topology that incorporates an extra dimension to encode Einstein's space-time as a latent manifold.

On the basis of evidence presented in this chapter, DM is postulated to have arisen during the creation of elementary particles in an early universe. In contrast with our present-day "flat" universe, this early universe was endowed with extreme geometric curvature and, consequently, was subject to special quantization rules. To validate this picture, an AI platform is leveraged in the guise of a quintessential autoencoder. This platform is capable of reverse engineering the "action principles" that underlie

DOI: 10.1201/9781003501534-5

the Standard Model of elementary particle physics. The deconstruction treats Einstein's 4D space-time as a "latent space" that gets decoded onto a 5-dimensional compact multiply connected manifold with no boundary. The manifold is endowed with an extra rolled-up dimension that spans a quark cross-section, which is the smallest material dimension, estimated to lie within the attometer (10^{-18} m) scale. It turns out that this compact fifth dimension stores stationary wave-matter with a rest mass that matches the vacuum expectation value of the Higgs boson. This result is pregnant with possibilities and implications and enabled us to estimate elementary particle masses with significant precision by an AI–based quintessential decoding of the elementary particle fields. The results point to the existence of an ur-Higgs boson in the early universe, specifically at the beginning of the "electroweak epoch", whose kinetic energy is not geometrically diluted along the standard 4D dimensions. This ur-Higgs and its heavier quantum relatives are identified by AI as DM. The results enable us to characterize DM vis-à-vis the geometric dilution of gravity shown to have taken place as the universe flattened and expanded to present-day levels.

The results obtained by AI and reported in this chapter may be summarized as follows:

- The universe topology is "revealed" by the CMB fluctuation spectrum. Unlike the Euclidean topology, this topology is compatible with the big bang scenario.

- AI can trace the origin of dark matter to the universe evolution by reverse engineering the Standard Model to incorporate a circular dormant dimension.

- On the dormant compact dimension with quark-size radius q, a de Broglie matter-wave has energy equal to the vacuum expectation value of the Higgs field. This is the ur-Higgs elementary particle, a good candidate for dark matter for the following reasons:

  - Incorporation of a fifth dimension is supported by evidence: SM elementary particle masses can be predicted through geometric dilution of the ur-Higgs.

  - The ur-Higgs formed in an early universe ($\sim 10^{-27}$ s after the big bang).

- It is massive and cold ($v \ll c$).

- It does not decay through communication with SM gauge bosons.

- It is only interactive with the SM via gravity.

- It endows other ur-particles with mass through ur-boson–conveyed geometric dilution.

On the other hand, DE is generated by quantum vacuum fluctuations exciting the quintessential field of the ur-Higgs and becoming stored as DM whenever the wavelength of the excitation fits a stationary wave condition along the dormant dimension.

Finally, a scenario is established as described by a space-time rendering of the universe evolution through geometric dilution of dark matter.

4.1 AI DECODES THE STANDARD MODEL

As shown in the scheme in Figure 2.12 and in Chapter 3, the concept of diagram commutativity is central to the encoding of a dynamical system into its latent simplified dynamics and, reciprocally, to the decoding of the latent dynamics onto a "quintessential" space-time that incorporates a dormant dimension [1]. This scheme is now adapted to decode the Standard Model (SM) onto a space with an extra dimension. Thus, the task of decoding the SM defined on the 4-dimensional space-time W/~, onto a 5-dimensional compact multiply connected manifold W requires a "quintessential" (i.e., "fifth essence") autoencoder operating in reverse. Typically, autoencoders simplify the dynamics to retain the latent coordinates [1]. In the case of interest to elementary particle physics, however, tangible observable processes are assumed to be taking place in a quotient space of equivalent classes modulo a dormant dimension, and the goal of the quintessential autoencoder is to decode the SM onto the space that incorporates the dormant fifth dimension (the "fifth essence").

This reverse quintessential autoencoder lifts the flow $\mathcal{F}: W/\sim \to W/\sim$ onto a flow $\tilde{\mathcal{F}}: W \to W$. The lifting is compatible if and only if it satisfies the commutativity relation: $\mathcal{F} \circ \pi = \pi \circ \tilde{\mathcal{F}}$, with $\pi: W \to W/\sim$ denoting the canonical projection that assigns each point in W to its equivalence class in W/~. Since we are introducing autoencoders for dynamical systems [1], it becomes imperative to cast the SM as a dynamical system. This prompts the question: How do we cast the flow $\mathcal{F} : W/\sim \longrightarrow W/\sim$ to represent a process described by the SM?

To answer the question, we may assimilate the interaction processes determined by the SM as transformations created by a time-dependent operator. This analogy is not complete, since the Lagrangian in the SM does not determine least action in a dynamic sense. For each elementary process, this operator has a time step τ associated with it, and this time step is precisely the lifetime of the gauge boson that communicates the force acting on the elementary particle and causing the elementary particle transformation [2]. Thus, the time step is obtained from the uncertainty principle: $\sim \dfrac{h}{2M_B c^2}$, where M_B is the rest mass of the relevant gauge boson. Thus, the autoencoder described in Chapter 3 will be operated in reverse in order to decode the SM by incorporating the dormant dimension shown to store gravity in a reverse engineering scheme. The NN for the quintessential autoencoder (QAE) is constructed in such a way that each input node represents the elementary particle field value (for simplicity assumed to be a complex number) in an equivalence class contained in $W/\sim$. The grid of nodes or equivalence classes has a mesh determining the level of resolution and each equivalence class is taken modulo the dormant coordinate, implying that it is represented by the four latent coordinates of the standard 4-dimensional space-time. Since the autoencoder is set up to be working in reverse, the output nodes represent the values of the decoded elementary particle field on points in the quintessential space W. Thus we have the following:

Lemma 4.1

The elementary particle field ϕ is decoded by the quintessential autoencoder as the ur-particle field $\tilde{\phi}$ if and only if $\exists \gamma_\phi$:

$$\phi \circ \pi = \gamma_\phi \circ \tilde{\phi}; \gamma_\phi : \mathbb{C} \to \mathbb{C} \tag{4.1}$$

The proof follows from the commutativity of the following diagram that would make the decoding compatible with the field defined on the 4-dimensional space-time:

$$\begin{array}{ccc} W & \to & W/\sim \\ \downarrow \tilde{\phi} & & \downarrow \phi \\ \mathbb{C} & \to & \mathbb{C} \end{array} \tag{4.2}$$

The field for an elementary particle with Lagrangian $\mathcal{L}$ satisfies the Euler-Lagrange equation:

$$\partial_\phi \mathcal{L} = \partial_\mu \frac{\partial \mathcal{L}}{\partial(\partial_\mu \phi)} \tag{4.3}$$

While the decoded field of the ur-particle obeys the relation:

$$\partial_{\tilde{\phi}} \tilde{\mathcal{L}} = \partial_\nu \frac{\partial \tilde{\mathcal{L}}}{\partial(\partial_\nu \tilde{\phi})}, \tag{4.4}$$

with its respective decoded Lagrangian $\tilde{\mathcal{L}}$ fulfilling the relation:

$$\int \tilde{\mathcal{L}}(\tilde{\phi}) dx dy_5 = \int \mathcal{L}(\phi) dx + \left[\int \left| \int (\tilde{\phi} - \phi) dx \right|^2 dy_5 \right]^{1/2} \tag{4.5}$$

Heretofore, the dormant coordinate is denoted y_5. ▪

Equation (4.5) is central to the process of quintessential decoding of the SM. It indicates that the field $\tilde{\phi}$ of the ur-particle is always the minimal smooth extension of the elementary particle field ϕ over the fifth coordinate, while the compactness of the latter coordinate ensures the convergence of the right term in the r.h.s. of Equation (4.5).

In principle, we do not and cannot *a priori* specify the radius r_0 of the circular dormant coordinate because this parameter, together with the geometric dilution parameter (or α pitch), must be determined in the optimization of the pair $(\tilde{\phi}, \gamma_\phi)$ in accord with the loss function $J(\alpha, r_0)$ associated to the diagram commutativity (Figure 4.1):

$$J(\alpha, r_0) = \sum_{y \in M_\alpha} \frac{1}{|M_\alpha|} \left\| (\phi \circ \pi) y - (\gamma_\phi \circ \tilde{\phi}) y \right\|^2 \tag{4.6}$$

$$(\alpha, r_0)(\phi) = \text{Arg Min } J \tag{4.7}$$

In practice, we shall see that the optimization dictated by the loss function attributes a unique geometric dilution value to each elementary particle in the SM, while r_0 remains constant and equal to the quark value q

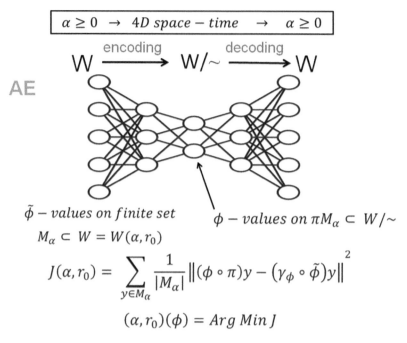

$$\boxed{\alpha \geq 0 \quad \rightarrow \quad 4D\ space - time \quad \rightarrow \quad \alpha \geq 0}$$

encoding   decoding

$$W \longrightarrow W/\sim \longrightarrow W$$

AE

$\tilde{\phi} - values\ on\ finite\ set$ $\phi - values\ on\ \pi M_\alpha \subset W/\sim$

$M_\alpha \subset W = W(\alpha, r_0)$

$$J(\alpha, r_0) = \sum_{y \in M_\alpha} \frac{1}{|M_\alpha|} \left\| (\phi \circ \pi)y - (\gamma_\phi \circ \tilde{\phi})y \right\|^2$$

$$(\alpha, r_0)(\phi) = Arg\ Min\ J$$

FIGURE 4.1   The quintessential autoencoder (AE) decoding a SM particle field ϕ on the 4-dimensional space-time that is regarded as the latent manifold $W/\sim$ of the quintessential space W. A loss function is parametrically determined by the pitch α and dormant dimension r_0, whose optimization yields the decoded field $\tilde{\phi}$.

in the attometer range [3]. Assuming and anticipating $r_0 = q$, the corresponding wave functions for ur-particles become parametrically dependent on geometric dilution.

By definition, the ur-Higgs boson mass is stored completely in the dormant dimension. Therefore, the wavefunction Y for the ur-Higgs boson is associated with $\alpha = 0$ (no geometric dilution) and its expectation energy becomes $\langle Y | \widetilde{\mathcal{H}_H} | Y \rangle = 246\ \text{GeV}$, where $\widetilde{\mathcal{H}_H}$ denotes the decoded Hamiltonian for the ur-Higgs boson. On the other hand, the wave function Ψ_H for the decoded Higgs boson satisfies $\langle \Psi_H | Y \rangle = \cos \alpha^*$, where $\alpha^* = 59°27' \approx \pi/3$. In general, for an ur-particle ζ with associated pitch α_ζ, we obtain $\langle \Psi_\zeta | Y \rangle = \cos \alpha_\zeta$. These observations are memoralized in Figure 4.2. The protocol for obtaining wave functions $\Psi : W \to \mathbb{C}$ for ur-particles decoding elementary particles is presented in Figure 4.3a,b.

It should be noticed that heavier relatives of the ur-Higgs boson are also generated in accord with the increase in the frequency $f = \dfrac{cn}{2\pi r_0} (n > 1)$

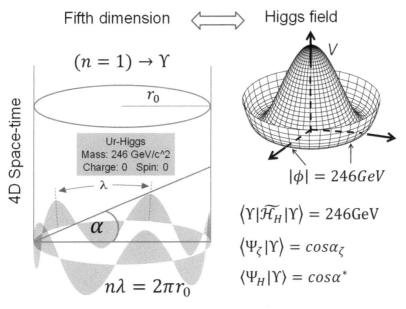

FIGURE 4.2 The dormant fifth dimension in the quintessential decoding of space-time. The fifth dimension represents the smallest material scale known, which corresponds to the quark cross-section, established to be in the attometer range [3]. The lowest energy stationary (de Broglie) matter-wave fully stored in the dormant dimension and named ur-Higgs has a rest mass energy corresponding to the vacuum expectation value of the Higgs boson field $|\phi| \approx 246$ GeV.

of the stationary wave along the dormant dimension. The space-time rendering of this quintessential particle hierarchy representing DM is obtained for integer values of n, as shown schematically in Figure 4.4. It reveals a ripple in the space-time cross-section, consistent with formation of a black hole as the mass M_n of the particle increases beyond the Planck mass: $M_n > M_P$. The signature of the black hole is the placement of its interior into the future, so time is forced to run backwards to escape the event horizon (Figure 4.4), a theoretical impossibility according to relativity.

The space-time rendering of the quantum field excitations representing the DM hierarchical tower becomes possible using the quantum gravity autoencoders introduced in Chapter 3. These AI systems can decode the quantum field excitations as relativistic space-time (Figure 4.5), showing a time reversal that is indicative of the interior of a black hole that emerges when n reaches the critical value n^* that yields the Planck mass (cf. Figure 4.4).

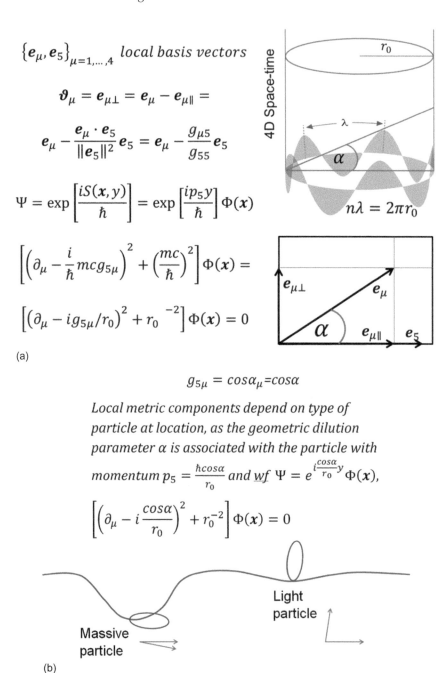

$$\{e_\mu, e_5\}_{\mu=1,\dots,4} \quad \textit{local basis vectors}$$

$$\vartheta_\mu = e_{\mu\perp} = e_\mu - e_{\mu\|} =$$

$$e_\mu - \frac{e_\mu \cdot e_5}{\|e_5\|^2} e_5 = e_\mu - \frac{g_{\mu5}}{g_{55}} e_5$$

$$\Psi = \exp\left[\frac{iS(x,y)}{\hbar}\right] = \exp\left[\frac{ip_5 y}{\hbar}\right] \Phi(x)$$

$$\left[\left(\partial_\mu - \frac{i}{\hbar} mc g_{5\mu}\right)^2 + \left(\frac{mc}{\hbar}\right)^2\right]\Phi(x) =$$

$$\left[\left(\partial_\mu - ig_{5\mu}/r_0\right)^2 + r_0^{-2}\right]\Phi(x) = 0$$

(a)

4D Space-time

r_0

λ

α

$n\lambda = 2\pi r_0$

$e_{\mu\perp}$

e_μ

α

$e_{\mu\|}$

e_5

$$g_{5\mu} = \cos\alpha_\mu = \cos\alpha$$

Local metric components depend on type of particle at location, as the geometric dilution parameter α is associated with the particle with

momentum $p_5 = \dfrac{\hbar\cos\alpha}{r_0}$ *and* <u>wf</u> $\Psi = e^{i\frac{\cos\alpha}{r_0}y}\Phi(x)$,

$$\left[\left(\partial_\mu - i\frac{\cos\alpha}{r_0}\right)^2 + r_0^{-2}\right]\Phi(x) = 0$$

Light
particle

Massive
particle

(b)

FIGURE 4.3 (a,b) Protocol to generate the wave functions for ur-particles decoding elementary particles in the SM.

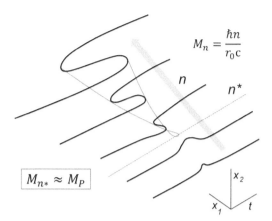

$$M_n = \frac{\hbar n}{r_0 c}$$

FIGURE 4.4 Space-time rendering of the quintessential hierarchy of relatives of the ur-Higgs boson obtained for $n > 1$. A cusp ripple in the space-time cross-section occurs consistently with formation of a black hole when the mass M_n of the particle increases beyond the Planck mass M_P, as $n^* \leq n$.

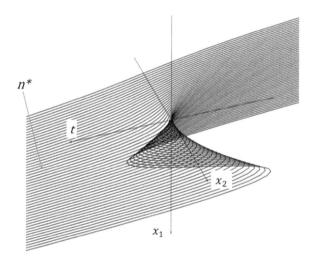

FIGURE 4.5 Quantum field excitations for the DM hierarchy decoded as relativistic space-time by a quantum gravity autoencoder.

4.2 REVERSE ENGINEERING OF THE STANDARD MODEL INTO A QUINTESSENTIAL MODEL INCLUDING DARK MATTER

In order to compute the quintessential decoding of an elementary particle field to generate the ur-particle field, the reverse autoencoder changes the metric of each Euclidean coordinate in W, compressing the metric

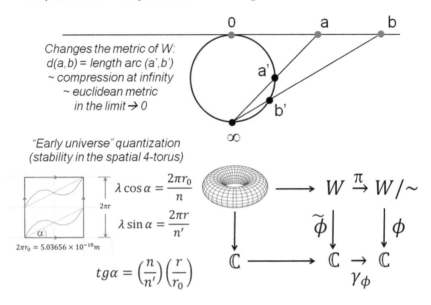

FIGURE 4.6 Early-universe toroidal decoding of the 4-dimensional space-time regarded as the latent space (W/~) for the quintessential space W homeomorphic to an "early-universe" 5-dimensional torus. The quintessential toroidal decoding of the particle field introduces a quantization required to avoid destructive interference on the 5-dimensional torus. The quantization of the ur-field is thus defined by four independent quantum numbers that are encoded into the standard particle attributes within the Standard Model (charge, mass, and spin) plus an extra attribute known as geometric dilution, only apparent in the quintessential decoded space W.

asymptotically as we approach infinity. This is a purely AI move, completely unsupervised and uninstructed and compactifies the space as it rolls up each Euclidean dimension onto a circle (Figure 4.6). The Euclidean metric on the line is changed into a new metric obtained by mapping a circle osculating the line, so that the distance between two points along the line is now evaluated as the length of the arc sustained between the projected points on the circle, as shown in Figure 4.6.

This new metric essentially maps W homeomorphically onto a 5-dimensional torus, so that states representing ur-particles become quantized in W with five quantum numbers due to the requirement to generate stationary de Broglie matter-waves in the 5-dimensional torus. Thus, quantization rules apply so that the wavelength λ for a elementary particle with pitch α must now fulfill the stationary-wave equation

$\lambda \cos \alpha = \dfrac{2\pi r_0}{n}$ for some quantum number (integer) n, and simultaneously fulfill four stationary-wave equations of the form $\lambda \sin \alpha = \dfrac{2\pi r}{n'}$ for quantum numbers n' for each coordinate in the 4-dimensional space-time endowed with the metric inherited from projection onto a circle of radius r (Figure 4.6). This implies that geometric dilution α, quantum number ratio $\left(\dfrac{n}{n'}\right)$, and aspect ratio $\left(\dfrac{r}{r_0}\right)$ are interrelated according to the four fundamental relations:

$$tg\alpha = \left(\frac{n}{n'}\right)\left(\frac{r}{r_0}\right) \tag{4.8}$$

For example, for the Higgs boson at $(n,n') = (1,1)$ with $\alpha = 59°27'$ ($tg\alpha = 1.69428$), we get the aspect ratio $\left(\dfrac{r}{r_0}\right) = 1.69428$. Since aspects ratios are invariants of the 5-dimensional torus and the pitch is fixed for each ur-particle, the set of five quantum numbers (n,n'_1,n'_2,n'_3,n'_4) determines the quintessential quantized decoding of each elementary particle in the SM. The decoding of each elementary particle field through the toroidal quantization of its ur-field in the quintessential space W is thus represented in Figure 4.6. The four elementary particle attributes in the SM – mass, geometric dilution, charge and spin – represent a specific encoding of the four independent quantum numbers in the toroidal quantization described by the four relations in Equation (4.8).

The geometric dilution associated with each elementary particle in the SM is thus obtained by computing the ur-field as the optimum quintessential decoding of the elementary particle field that minimizes the loss function in fulfillment of Eqs. (4.6) and (4.7). This decoded field minimizes the quintessential Lagrangian defined by Equation (4.5). The results are presented in Figure 4.7.

The decoding of the SM is fully implemented by ensuring through parametric optimization the commutativity of the dual action diagram presented in Figure 4.8. To introduce this higher level of description some notation becomes necessary: The space of smooth scalar fields for ur-particles is $C^2(W, \mathbb{C})$, while the space $C^2(W/\sim, \mathbb{C})$ contains the elementary particle scalar fields in the SM. The respective dual spaces $C^2(W, \mathbb{C})^*$ and $C^2(W/\sim, \mathbb{C})^*$ contain the actions defined by functionals-actions of the type: $\tilde{\mathfrak{A}}\left(\tilde{\mathcal{L}}\right) = \int \tilde{\mathcal{L}}\left(\tilde{\phi}, \partial_\nu \tilde{\phi}\right) dy^\nu$ and $\mathfrak{A}\left(\mathcal{L}\right) = \int \mathcal{L}\left(\phi, \partial_\mu \phi\right) dx^\mu$, respectively. As described in Chapter 2, an elementary particle Lagrangian $\mathcal{L}$ becomes

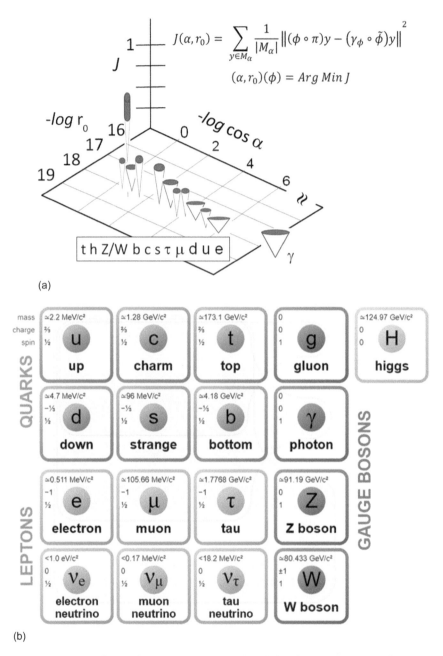

$$J(\alpha, r_0) = \sum_{y \in M_\alpha} \frac{1}{|M_\alpha|} \left\| (\phi \circ \pi) y - (\gamma_\phi \circ \tilde{\phi}) y \right\|^2$$

$$(\alpha, r_0)(\phi) = Arg\ Min\ J$$

(a)

(b)

FIGURE 4.7 Pitch and dormant dimension (α, r_0) for the optimization of the loss function $J(\alpha, r_0)$ for the quintessential autoencoder that decodes each elementary particle field in the Standard Model. (a) Loss function optimization in (J, α, r_0)-space for each elementary particle denoted following notation in (b). (b) Table of elementary particles in the Standard Model.

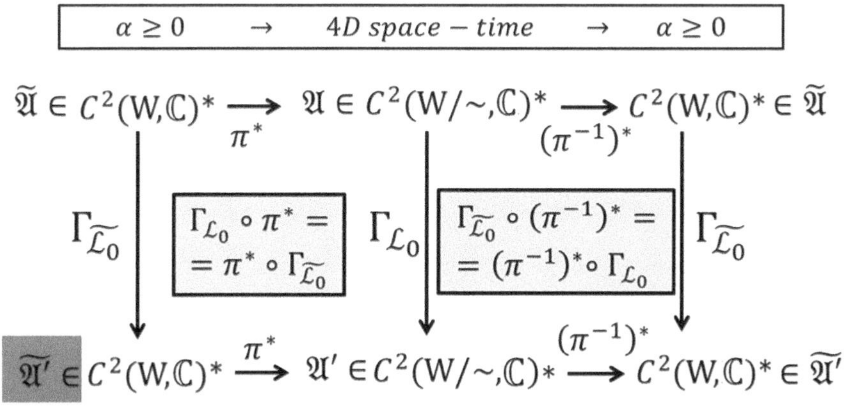

$$\boxed{\alpha \geq 0 \qquad \rightarrow \qquad 4D\ space-time \qquad \rightarrow \qquad \alpha \geq 0}$$

$$\widetilde{\mathfrak{A}} \in C^2(W,\mathbb{C})^* \xrightarrow{\ \pi^*\ } \mathfrak{A} \in C^2(W/{\sim},\mathbb{C})^* \xrightarrow{\ (\pi^{-1})^*\ } C^2(W,\mathbb{C})^* \in \widetilde{\mathfrak{A}}$$

$$\Gamma_{\widetilde{\mathcal{L}_0}} \quad \boxed{\begin{array}{c}\Gamma_{\mathcal{L}_0} \circ \pi^* = \\ = \pi^* \circ \Gamma_{\widetilde{\mathcal{L}_0}}\end{array}} \quad \Gamma_{\mathcal{L}_0} \quad \boxed{\begin{array}{c}\Gamma_{\widetilde{\mathcal{L}_0}} \circ (\pi^{-1})^* = \\ = (\pi^{-1})^* \circ \Gamma_{\mathcal{L}_0}\end{array}} \quad \Gamma_{\widetilde{\mathcal{L}_0}}$$

$$\widetilde{\mathfrak{A}}' \in C^2(W,\mathbb{C})^* \xrightarrow{\ \pi^*\ } \mathfrak{A}' \in C^2(W/{\sim},\mathbb{C})^* \xrightarrow{\ (\pi^{-1})^*\ } C^2(W,\mathbb{C})^* \in \widetilde{\mathfrak{A}}'$$

$$\mathfrak{A} = \mathfrak{A}(\mathcal{L}) = \int \mathcal{L}(\phi, \partial_\mu \phi) dx^\mu \qquad \widetilde{\mathfrak{A}} = \int \widetilde{\mathcal{L}}(\tilde{\phi}, \partial_\nu \tilde{\phi}) dy^\nu$$

$$\Gamma_{\mathcal{L}_0}(\mathfrak{A}) = \mathfrak{A}' = \mathfrak{A}(\mathcal{L}')\ where$$

$$\int [\mathcal{L} + \mathcal{L}_0] dx = \int [\mathcal{L}' + \mathcal{L}'_0] dx'$$

FIGURE 4.8 Quintessential decoding of the Standard Model implemented by ensuring the commutativity of the action diagram through parametric optimization. In this decoding context dual to that represented in Figure 6.1, the gauge boson with Lagrangian $\mathcal{L}_0$ is represented as a map $\Gamma_{\mathcal{L}_0} : C^2\left(W/{\sim},\mathbb{C}\right)^* \to C^2\left(W/{\sim},\mathbb{C}\right)^*$ defined over action space as $\Gamma_{\mathcal{L}_0}\mathfrak{A}(\mathcal{L}) = \mathfrak{A}(\mathcal{L}')$. Similarly, the corresponding decoded ur-boson is represented by the map $\Gamma_{\widetilde{\mathcal{L}}} : C^2\left(W,\mathbb{C}\right)^* \to C^2\left(W,\mathbb{C}\right)^*$ defined as $\Gamma_{\widetilde{\mathcal{L}_0}}\widetilde{\mathfrak{A}}\left(\widetilde{\mathcal{L}}\right) = \widetilde{\mathfrak{A}}\left(\widetilde{\mathcal{L}}'\right)$. So the consistent decoding of the SM requires that the following commutativity relation is valid for any generic boson with Lagrangian $\mathcal{L}_0 : \Gamma_{\mathcal{L}_0} \circ \pi^* = \pi^* \circ \Gamma_{\widetilde{\mathcal{L}_0}}$, where π^* is the dual map of the canonical projection of the quintessential space W onto the quotient (latent) space W/~.

transformed into another elementary particle Lagrangian $\mathcal{L}'$ when the first elementary particle interacts with another elementary particle (gauge boson) with Lagrangian $\mathcal{L}_0$. These interactions define the flow in action space that enables the SM to be represented as a dynamical system with propagator time step defined as the lifetime of the gauge boson with associated Lagrangian $\mathcal{L}_0$ that communicates the force acting on the elementary particle and induces its transformation. Thus, the SM may be decoded at the action level using dual autoencoders adapted to dynamical systems of the type described in Chapter 3 but acting in reverse, where the 4-dimensional space-time represents the latent space.

In the dual context of this decoding, the gauge boson with Lagrangian $\mathcal{L}_0$ is represented as a map $\Gamma_{\mathcal{L}_0} : C^2\left(W/{\sim},\mathbb{C}\right)^* \to C^2\left(W/{\sim},\mathbb{C}\right)^*$ defined as

$\Gamma_{\mathcal{L}_0}\mathfrak{A}(\mathcal{L})=\mathfrak{A}(\mathcal{L}')$. Similarly, the corresponding ur-boson is represented in this dual space by the map $\Gamma_{\widetilde{\mathcal{L}_0}}:C^2(W,\mathbb{C})^* \to C^2(W,\mathbb{C})^*$ defined as $\Gamma_{\widetilde{\mathcal{L}_0}}\widetilde{\mathfrak{A}}(\widetilde{\mathcal{L}})=\widetilde{\mathfrak{A}}(\widetilde{\mathcal{L}'})$. So the decoding of the SM becomes operational when it is also validated at the level of actions, implying that the following commutativity relation is valid for any gauge boson with generic Lagrangian denoted $\mathcal{L}_0$:

$$\Gamma_{\mathcal{L}_0}\circ\pi^* = \pi^*\circ\Gamma_{\widetilde{\mathcal{L}_0}} \tag{4.9}$$

In Equation (4.9), we assume that the canonical projection $\pi : W \to W/\sim$ induces the dual canonical projection $\pi^* : C^2(W, \mathbb{C})^* \to C^2(W/\sim, \mathbb{C})^*$ mapping the action space for W onto the action space for the SM.

Geometric dilution parametrized by the pitch angle α may be communicated gravitationally in the quintessential space W through the respective ur-boson$_\alpha$, as schematically represented in the Feynman diagram [2] displayed in Figure 4.9. Thus, the ur-Higgs may decay into a geometrically diluted ur-particle with energy $E_\alpha = \left(\dfrac{c\hbar}{r_0}\right)\cos\alpha$ and thereby transfer the geometrically diluted mass m (expressed in GeV/c²), satisfying the equation cos $\alpha = (246\text{-}m)/246$. The mass is captured by an ur-particle with rest mass m' that gets transformed into an ur-particle with mass $m + m'$ upon interaction with the ur-boson α that has been emitted or interacted with the ur-Higgs boson.

On the other hand, ur-bosons are predicted by the quintessential autoencoder to participate in elementary processes in the vacuum involving quark-pair annihilation with lepton-pair decay, as described in Figure 4.10. In this case, the ur-boson has a geometric dilution $=-\log\cos\alpha =-\log\left(\dfrac{c^2 m_q}{246\text{GeV}}\right)$, where m_q is the quark rest mass.

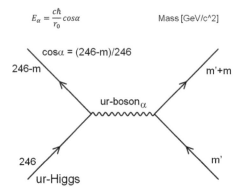

FIGURE 4.9 Communication of geometric dilution in the quintessential space W represented as a Feynman diagram (cf. [2]), involving an ur-particle with mass m', the ur-Higgs boson and an ur-gauge boson associated with the pitch value α.

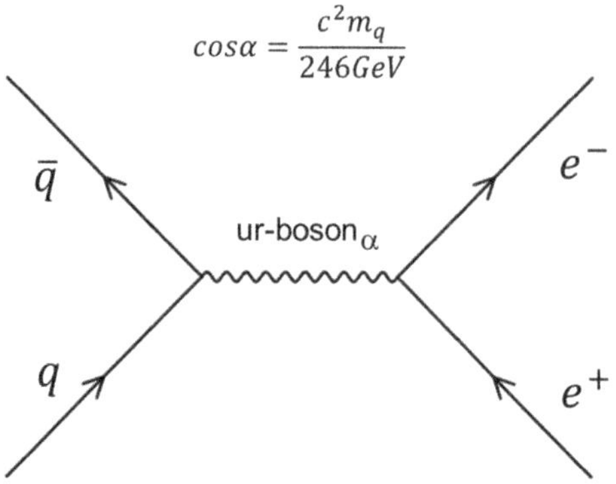

$$\cos\alpha = \frac{c^2 m_q}{246GeV}$$

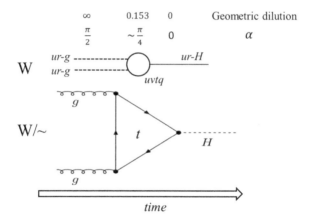

FIGURE 4.10   Predicted quark pair annihilation mediated by an ur-boson whose geometric dilution is determined by the quark's rest mass.

FIGURE 4.11   Quintessential lifting of the elementary process of Higgs boson creation through fusion of a pair of gluons mediated by a virtual top quark.

Furthermore, the deconstruction of the SM enables the quintessential lifting of any elementary process. This is illustrated in Figure 4.11, where the generation of the Higgs boson through a gluon pair fusion is quintessentially lifted to a process where communication with the ur-counterpart of the virtual top quark (uvtq) triggers the drop in geometric dilution required to render the ur-Higgs boson.

4.3   EXPERIMENTAL VALIDATION OF THE QUINTESSENTIAL DECODING OF THE STANDARD MODEL

The reverse quintessential autoencoder decodes the elementary particle fields in the SM as ur-particles with masses determined by the geometric dilution of the ur-Higgs stationary wave stored on the dormant fifth coordinate. The geometric dilution may be directly determined from the pitch of the wave front associated with the quintessentially decoded wave function for the ur-particle. Wave functions for elementary particles are decoded as maps $\Psi : W \to \mathbb{C}$ following the same tenets that apply to the decoding of particle fields (Section 4.2). The decoding of the photon (γ) has infinite dilution ($v = -\log \cos \alpha = \infty$, $\alpha = \pi/2$), yielding a zero mass (Figure 4.7a), and hence this gauge boson communicating the electromagnetic force travels at the speed of light. Thus, the wavefunction $\Psi_\gamma : W \to \mathbb{C}$ for the ur-photon is orthogonal to that of the ur-Higgs $\langle \Psi_\gamma | \Upsilon \rangle = \cos(\pi/2) = 0$. At the opposite end of the spectrum, the heaviest elementary particle, the top quark (t), gets decoded into the ur-t which has the lowest geometric dilution of all decoded elementary particles in the SM: $v = -\log \cos \alpha = 0.153$, $\alpha \approx \pi/4$. This yields $\langle \Psi_t | \Upsilon \rangle = 0.703$, a good approximation to the ratio between elementary particle mass and Higgs vacuum expectation value (v): $\dfrac{m}{v} = \dfrac{173.1}{246} \approx \langle \Psi_t | \Upsilon \rangle$. The following relation is valid for all gauge bosons in the SM:

$$\frac{m_\zeta}{v} \approx \langle \Psi_\zeta | \Upsilon \rangle \tag{4.10}$$

In the case of fermions, the relation becomes

$$g_\zeta \approx \langle \Psi_\zeta | \Upsilon \rangle, \tag{4.11}$$

where g_ζ is the respective Yukawa coupling constant [2]. The Yukawa coupling to the Higgs field can thus be interpreted as an effective geometric dilution parameter for ur-fermions in quintessential space as it follows from the relation $g = \cos \alpha$.

A linear plotting $\dfrac{m_\zeta}{v}, g_\zeta - \langle \Psi_\zeta | \Upsilon \rangle$ yields $R^2 = 0.88$ as shown in Figure 4.12 for all elementary particles in the SM. The resulting prediction of masses is based on the geometric dilution that optimizes the loss function $J(\alpha, q)$ for the decoding of each particle field into its ur-field. The implication is that geometric dilution, or equivalently, the pitch angle α, is a fundamental parameter that determines the mass of ur-particles originally enshrined in the ur-Higgs. When encoded into the quotient (latent) space $W/\sim$, this relatively simple process becomes the much more complex process by which mass is endowed by the Higgs boson in its interplay with the

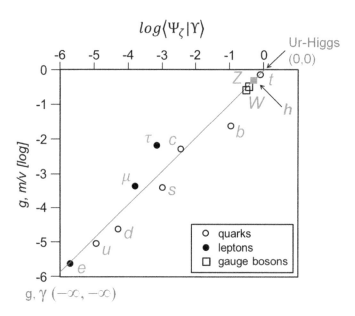

FIGURE 4.12 Prediction of the mass for each elementary particle (ζ) in the Standard Model through the correlation $\frac{m_\zeta}{v}, g_\zeta - \langle \Psi_\zeta \, | \, \Upsilon \rangle$ ($R^2 = 0.88$), where $\frac{m_\zeta}{v}$ is the ratio of gauge boson mass over vacuum expectation value for the Higgs boson and g_ζ is the Yukawa coupling parameter for the case when ζ denotes a fermion. The linear correlation validates the quintessential decoding of the Standard Model as reverse engineering.

particle fields [2]. The commutativity of the following diagram represents the decoding of the Higgs mass-endowment process:

$$W \xrightarrow{\pi} W/\sim$$
$$\text{geometric dilution} \downarrow \widetilde{\mathfrak{F}}_H \quad \downarrow \mathfrak{F}_H \, \text{Higgs mechanism}$$
$$W \xrightarrow{\pi} W/\sim$$

$$(4.12)$$

The flow $\mathfrak{F}_H$ represents the endowment of mass through interaction with the Higgs boson. This process should be interpreted in the sense that for a generic particle with Lagrangian $\mathcal{L}$, the following relation holds: $\Gamma_{\mathcal{L}_H} \mathfrak{A}(\mathcal{L}) = \mathfrak{A}(\mathcal{L}')$, where the massless original Lagrangian $\mathcal{L}$ has changed to $\mathcal{L}'$, now endowed with mass conferred though interaction with the Higgs boson, while the latter becomes a Nambu-Goldstone massless boson [2]. This process is decoded at the W-level by the flow $\widetilde{\mathfrak{F}}_H$, representing the endowment of mass on the ur-particle through geometric dilution of the ur-Higgs boson (Figure 4.9). In other words, the following relation holds at the encoded W-level: $\Gamma_{\widetilde{\mathcal{L}_H}}\widetilde{\mathfrak{A}}\left(\widetilde{\mathcal{L}}\right) = \widetilde{\mathfrak{A}}\left(\widetilde{\mathcal{L}'}\right)$, where $\widetilde{\mathcal{L}_H}$ is the Lagrangian for the ur-Higgs boson, and $\widetilde{\mathcal{L}}$, $\widetilde{\mathcal{L}'}$ are the decoded Lagrangians for the massless ur-particle and for the ur-particle endowed with mass through the geometric dilution determined by the operator $\Gamma_{\widetilde{\mathcal{L}_H}}$, as illustrated in the Feynman diagram of Figure 4.9.

4.4 GEOMETRIC DILUTION PARAMETER AS A PROXY FOR EXTRINSIC TIME IN THE EVOLUTION OF THE UNIVERSE

A straightforward application of Heisenberg's uncertainty principle reveals that the emergence of the ur-Higgs, the primeval ur-particle implicated in baryogenesis, can be traced to a very specific time in the evolution of the universe after the big bang: $\tau \approx \dfrac{r_0}{2c} = 1.337 \times 10^{-27}$ s. This takes us to the so-called electroweak epoch when electromagnetism and the weak nuclear force remained merged into the so-called electroweak force [4]. At energies in the order of the vacuum expectation energy for the Higgs field (246 GeV), this merging remained energetically above the critical bound required for symmetry breaking, estimated in the SM at 159.5 GeV.

The flattening of the universe begins with the expansion driven by quintessential geometric dilution with $\alpha \neq 0$ (Figure 4.13), so gravity becomes less of an attribute of intrinsic geometry with extreme curvature [5] and more an attribute of the energy stored as mass in geometrically diluted incarnations of the ur-Higgs. The timing of universe evolution is in fact dictated by geometric dilution through the quantum relation

$$\tau \approx \frac{r_0}{2c}\left(\frac{1}{\cos\alpha}\right) = 1.337 \times 10^{-27}\left(\frac{1}{\cos\alpha}\right) s \qquad (4.13)$$

The relation (4.13) marks the timing of creation of an ur-particle with mass $m = \dfrac{\hbar\cos\alpha}{cr_0}$.

Thus, the emergence of the decoded Higgs boson (Figure 4.14) can be traced to the time

$$\tau \approx 1.337 \times 10^{-27}\left(\frac{1}{\cos\alpha^*}\right) s = 2.632 \times 10^{-27} \qquad (4.14)$$

This time marks the end of the electroweak epoch and is followed by the emergence of the decoded gauge bosons Z and W, which signal the

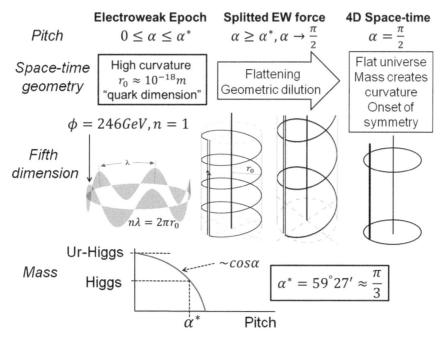

FIGURE 4.13 Flattening of the universe as a measure of time starting at the big bang event. The universe expansion is driven by quintessential geometric dilution with $\alpha > 0$, so gravity becomes less of an attribute of intrinsic geometry of the quintessential space and more an attribute of the energy stored as mass in geometrically dilutions of the ur-Higgs boson. The timing t in universe evolution starting at the big bang becomes $t \approx \dfrac{r_0}{2c}\left(\dfrac{1}{\cos\alpha}\right)$.

splitting of the electroweak force, as these bosons communicate the weak nuclear force when encoded in the latent space-time manifold W/~. All gauge bosons are created through vacuum fluctuation and stored into the diluted dormant dimension at dilution with cos α=0.32 (Figure 4.12), which corresponds to the electroweak splitting time $\tau \approx 4.178 \times 10^{-27}$s. The ur-fermions that emerge after this time can decay as the gauge bosons have already stepped into existence and hence convey the interaction required (Figure 4.14).

At this point, the quintessential space W is still endowed with significant local curvature and will become flat only asymptotically at infinite geometric dilution ($v \to \infty$) corresponding to the limit $\alpha \to \dfrac{\pi}{2}$. This is the limit for the emergence of the photon (γ), the newest and massless particle. Thus, the advent of the photon becomes indicative of a quasi-flat universe resulting from infinite geometric dilution of the dormant dimension (Figure 4.14).

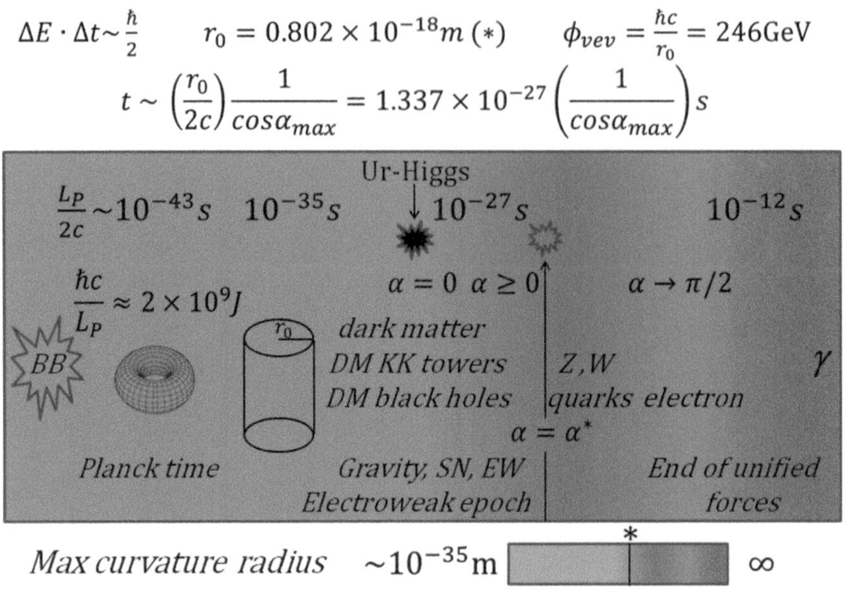

$$\Delta E \cdot \Delta t \sim \frac{\hbar}{2} \qquad r_0 = 0.802 \times 10^{-18} m \ (*) \qquad \phi_{vev} = \frac{\hbar c}{r_0} = 246 \text{GeV}$$

$$t \sim \left(\frac{r_0}{2c}\right)\frac{1}{cos\alpha_{max}} = 1.337 \times 10^{-27} \left(\frac{1}{cos\alpha_{max}}\right)s$$

FIGURE 4.14   Universe timeline as quintessential geometric dilution.

It is an undisputed fact that the photon already exists in our universe, and on the other hand, the big bang happened approximately 13.8 billion years ago; therefore, the flatness of the universe should be considered to be only approximate, but it is indeed a strikingly good approximation. The current pitch $\alpha = \alpha(\tau_u)$, corresponding to the age of the universe estimated at $\tau_u \sim 4.35 \times 10^{17}$s can be calculated effectively as

$$\alpha = \text{arc} \cos\left(\frac{r_0}{2c\tau_u}\right) = \text{arc} \cos\left(3.1 \times 10^{-45}\right) \cong \text{arc} \cos 0 = \frac{\pi}{2} \qquad (4.15)$$

It should be noted also that the ur-photon and the photon itself constitute for all purposes one and the same entity, since there is no projection onto the dormant dimension. Thus, the present day represents a singularity in which the dormant dimension may be set to be zero, and therefore, quotient space merges and identifies with underlying quintessential space. But this implies that the vacuum expectation value ϕ_{vev} for the Higgs field ($\sim \hbar c/r_0$) will ultimately be infinite, and that the current value at 246 Gev that sustains the universe as we know it represents a metastable state. We do not know when the current ϕ_{vev} for the Higgs field will decay to infinity, as that would depend on the energy barrier that must be overcome

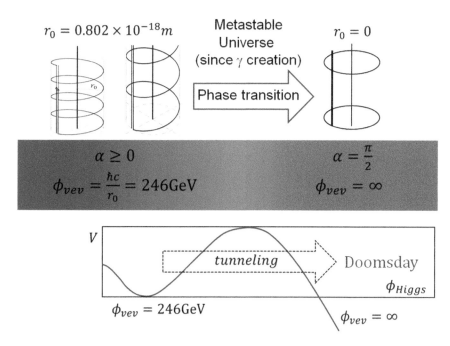

$$r_0 = 0.802 \times 10^{-18} m \qquad \text{Metastable Universe (since } \gamma \text{ creation)} \qquad r_0 = 0$$

Phase transition

$$\alpha \geq 0 \qquad\qquad \alpha = \frac{\pi}{2}$$

$$\phi_{vev} = \frac{\hbar c}{r_0} = 246 \text{GeV} \qquad \phi_{vev} = \infty$$

V

tunneling → Doomsday

ϕ_{Higgs}

$$\phi_{vev} = 246 \text{GeV} \qquad \phi_{vev} = \infty$$

FIGURE 4.15 Catastrophe scenario associated with the metastability of the Higgs boson in the present-day universe at nearly infinite geometric dilution.

through quantum tunneling. But we can be certain that the universe, as we know it, is metastable [4, 5], hence prone to undergo a phase transition at some point (Figure 4.15). When that phase transition materializes, a big crunch will take place as every particle will be endowed with infinite mass, taking the universe back to the starting point in a big bang scenario.

4.5 DARK MATTER IDENTIFIED BY QUINTESSENTIAL DECODING OF THE STANDARD MODEL

The conclusions drawn so far from the discussion in the preceding sections may be summarized as follows:

The universe topology is inferred through an argument of Parmenides' type (Chapter 2) and "revealed" by the CMB fluctuation spectrum, and it is identified with a four-dimensional torus, corresponding to a quasi-flat multiply connected compact space with no boundary.

This topology is compatible with the big bang scenario for universe evolution [4], which admits dimensional expansion with retention of compactness but not a change in topology, as the latter introduces a disruption in the fabric of space-time that is forbidden by general relativity [5].

The toroidal topology admits a circular dormant dimension as a compact extra dimension, whereas a Euclidean space cannot incorporate this compact dimension without a change in topology along the universe evolution.

AI can trace the origin of DM in the universe evolution, since on a compact extra dimension with quark-size radius q [3], a de Broglie wave stores energy equal to the vacuum expectation value of the Higgs field.

This is the ur-Higgs particle, an excitation of the quintessence field that has been identified, together with its hierarchical tower, as DM [6] by AI because:

- Incorporation of a fifth dimension is supported by evidence: elementary particle masses can be accurately predicted through the geometric dilution of the ur-Higgs quintessence field that is determined from the decoding of the particle fields (Figure 4.7).

- The ur-Higgs particle formed during the electroweak epoch in an early universe ($\sim 10^{-27}$ s after the big bang).

- It is massive (246 GeV/c^2) and cold (speed $\ll$ c).

- It does not decay through communication with SM gauge bosons.

- It is only interactive with the SM via gravity.

- It endows other ur-particles with mass through ur-boson-conveyed geometric dilution.

Much has been written about the difficulty in identifying the physical underpinnings of dark energy (DE) as the energy associated with quantum vacuum fluctuations (particles popping in and out of existence) [3]. In principle, the DE density required to observe the measured universe expansion is estimated at $\rho_{DE} \approx 5 \times 10^{-10}$ Jm^{-3}. This value is in stark disagreement with naïve calculations of the vacuum energy density, yielding an estimated DE density with a colossal discrepancy amounting to a staggering 120 orders of magnitude compared with the experimentally obtained figure. In cosmology, this problem is often referred to as the cosmological constant problem or vacuum catastrophe [5]. The so-called cosmological constant [4] is often regarded as the default model for DE, whereby the geodesic fabric of space-time has a constant non-zero energy density that yields an antigravitational pull background [5]. The problem essentially describes the disagreement between the observed values of vacuum energy density (the small value of the cosmological constant) and theoretical large value of zero-point energy that is obtained from quantum field theory. As said, the quantum vacuum energy contribution to the cosmological constant is calculated to be as much as 120 orders of magnitude greater than the one observed, a calamitous state of affairs that should in all likelihood be referred to as the largest disagreement between theory and experiment in the history of physics. This prompted scientists to explore other options for what DE might be.

Much of the difficulty and controversy evaporates as the extra compact dimension of quark-like material scale is incorporated in the quintessential decoding of the latent manifold representing standard space-time. This is because the excitation of the quintessential ur-Higgs field through quantum vacuum fluctuations enables the storage of DE as DM. In other words, the AI–enabled autoencoder technology dictates that we cannot conceive DE independently of DM and the dichotomy becomes a consequence of the quintessential decoding of our universe through the incorporation of the dormant fifth dimension.

To distill the thrust of this discussion we may state that the decoding of the SM by a quintessential autoencoder yields an ur-field that is not part of the SM. This quintessential field serves as the excitation vehicle for DE. In turn, the DE gets occasionally stored as DM for wavelengths that fit a stationary wave condition along the dormant fifth dimension. Thus, AI shows that DE and DM become phenomenologically tied up to the quintessential space, while becoming opaque to our understanding when regarded within the standard 4-dimensional space-time. Only when the latter is regarded as a latent space, the true nature of DM and DE may be revealed.

4.6 DARK MATTER AND DARK ENERGY HOLOGRAPHICALLY DECODED BY ADOPTING GEOMETRIC DILUTION AS A PROXY FOR TIME

In Chapter 3 the challenging problem of quantum gravity was addressed by constructing a learning system with stochastic connectivities and hidden variables where gravity and quantum behavior become emergent properties in a statistical physics scheme for machine learning. The thrust was to implement an AI–based version of the universe as a quantum holographic autoencoder. This system operates under the tenet that the emergent quantum behavior arises in a neural network equilibrated on the nontrainable hidden variables upon which a relativistic string gravitational scheme may be constructed. Conversely, in the nonequilibrium regime prior to equilibration of nontrainable variables, the network is endowed with emergent gravity. The behavior of the neural network is examined in the limits where the bias vector, weight matrix, and state vector of neurons can be modeled as stochastic variables that undergo a learning evolution. These dynamics are described by a time-dependent Schrödinger equation compatible with the relativistic decoding enshrined in the commutativity of the diagram in Figure 3.20. The results illustrate the emergence of quantum mechanics and general relativity in neural networks governed by a statistical physics scheme that can operate in two different thermodynamic regimes. In this way, the quantum metamodel for gravity fulfills at least in part a major imperative for physicists seeking a unified field theory.

The holographic autoencoder is now set up to adopt the toroidal topology of space-time described in Figure 4.6 with the geometric dilution parameter $v = -\log \cos \alpha$ as the proxy for time. With the default parameters given in Chapter 3, the timing of events follows exactly the description in Figure 4.14, with the advent of dark matter at $\sim 10^{-27}$s. As dictated by the holographic autoencoder, the complete materialization of the universe and its evolution within the big bang scenario can be realized through the geometric dilution of gravity, originally decoded as a stationary wave in the quintessential 5-dimensional torus.

4.7 DARK MATTER AND DARK ENERGY ARE NOT IDENTIFIED IN STANDARD EXTENSIONS OF THE STANDARD MODEL

The concept of reverse autoencoder, or rather, "autodecoder", has featured profusely in the previous discussion on the AI–enabled reverse engineering of the SM. In time-series data science, the autoencoder is regarded as the time-series distiller that generates the essential or latent model in a dynamical system [1]. In our context of interest, we assume the SM is in and of itself the latent model that entrains the full dynamics and that the leveraging of reverse autoencoder technology serves the purpose of determining the latter.

As a specialized learning system, the autoencoder typically needs to be trained to decode the latent dynamics. In the particular case of interest, the training does not require *a priori* knowledge of quintessential time series. This is because we have a generic way of deriving the quintessential Lagrangian

$\tilde{\mathcal{L}}$ that decodes a particle Lagrangian $\mathcal{L}$. Thus, the decoded Lagrangian satisfies the relation: $\int \tilde{\mathcal{L}}(\tilde{\phi}) dx dy_5 = \int \mathcal{L}(\phi) dx + \left[\int \left| \int (\tilde{\phi} - \phi) dx \right|^2 dy_5 \right]^{1/2}$. By applying this relation to a given pair $(\phi, \mathcal{L})$ representing a particle in the SM, the dynamics encoded in the SM is fleshed out onto the quintessential space (a compact and locally flat multiply connected manifold). This simply requires the decoding each particle field $\phi \to \tilde{\phi}$ so as to optimize (minimize) the term $\int \tilde{\mathcal{L}}(\tilde{\phi}) dx dy_5$.

To deploy autoencoder technology, we have turned the SM into a dynamical system by setting the time step for an elementary process equal to the lifetime of the (ephemeral) boson that mediates the particle transformation. Evidently, the center manifold entrainment [7–9] or subordination of the quintessential dynamical system to the SM leaves out dark energy and dark matter, so we cannot assume that the SM, staggeringly successful as it is, provides the complete description of the elementary processes that take place in the universe.

Our discussion and analysis in this chapter has focused primarily on elucidating what is left out with the reduction or entrainment of the quintessential dynamics by the SM, and we have found a striking answer: DM and DE. This is at some level disconcerting because both clearly have a very significant bearing on the observed dynamics of deep space. To be specific, they constitute over 95% of the gravitational budget of the detectable universe at large scales. Furthermore, this clearly introduces a contradiction: the existence of DM and DE implies that the SM cannot entrain or subordinate the universe as a dynamical system that materializes in the quintessential space. This contradiction can only be properly accounted for in one of two ways: 1) the SM does not represent the latent dynamics of the universe, or 2) the quintessential dynamics is in fact irreducible: it does not admit a latent dynamical system. This leads to a paradox since any of the two alternatives implies that an autoencoder could not have been *a priori* used to elucidate the quintessential dynamics spanned by the latent dynamics enshrined in the SM. We know this to be wrong, since an autoencoder has indeed been used as decoder of the SM.

This fundamental paradox does not undermine the AI enablement of the reverse engineering of the SM because it is based on the obviously incorrect assumption that DM and DE can be obtained from extensions of the SM. This means that DM and DE are inherently quintessential, which was from the start a basic tenet in the AI approach put forth in this book. The fundamental implication is that the decoding of the SM is insufficient to account for the quintessence fields that realize DM and DE when excited along the dormant dimension.

Notwithstanding the previous conclusions, experimental efforts engaging multinational consortia, like the Large Hadron Colider, or deploying underground detectors are still well underway to detect DE and DM as extensions of the SM. Such efforts have been, not surprisingly, unsuccessful. It seems that for all that we revere the Copernican revolution, we still like to see ourselves as the center of the universe, and AI may be teaching us a lesson in that regard....

4.8 THE UNIVERSE WAS CREATED BY GEOMETRIC DILUTION OF DARK MATTER

As previously discussed, the storage of energy along the dormant dimension with radius r_0 generates a hierarchical tower associated with the ur-Higgs boson ($n = 1$; mass = 246 GeV/c²). The relatives of this particle for $n > 1$ have masses $M_n = \dfrac{\hbar n}{r_0 c}$. A critical accretion of this DM is indexed by $n = n^*$ satisfying the relation:

$$n^* = \min n : n \geq r_0 \sqrt{\frac{c^3}{\hbar G}} \tag{4.16}$$

Thus, the critical mass M_{n^*} becomes commensurable with the Planck mass M_P. This implies that we get a black hole when $n > n^*$. To understand the evolution of this enormous concentration of DM, we need to define "time outside space-time," that is, beyond the local relativistic time, t_{s-t}, that serves as a local dimension for space-time (cf. Figure 4.16). While spanning a far larger scale, the extrinsic time should be identifiable with t_{s-t} at the locus. We know that a proxy for evolutionary time is furnished by geometric dilution, hence, as shown in the space-time rendering of the hierarchical tower of DM, the $n > n^*$ ur-particles will eventually dilute, thus removing the space-time ripple that causes time reversal (Figure 4.16). This picture describes the evolution of the universe from DM, identifying geometric dilution of the excitation along the dormant coordinate as the mechanistic agent promoting particle creation concurrently with the progressive removal of the space-time ripple.

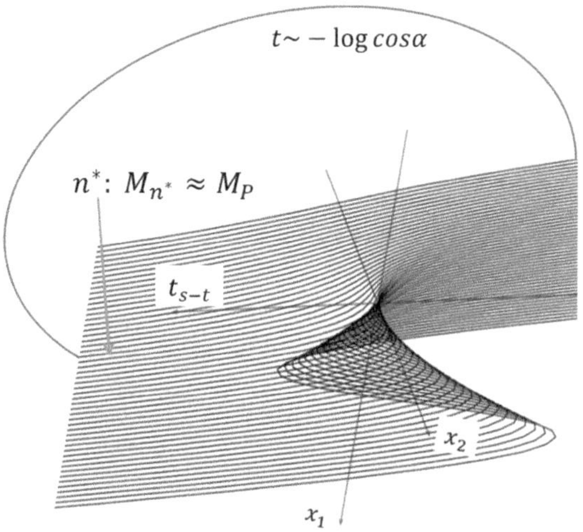

$$t \sim -\log \cos \alpha$$

$$n^*: M_{n^*} \approx M_P$$

$$t_{s-t}$$

$$x_2$$

$$x_1$$

FIGURE 4.16   Space-time rendering of the evolution of the universe by geometric dilution of dark matter.

REFERENCES

1. Fernández A (2022) *Topological Dynamics for Metamodel Discovery with Artificial Intelligence.* Chapman & Hall/CRC, Taylor and Francis, London, UK.
2. Feynman RP, Weinberg S (1999) *Elementary Particles and the Laws of Physics.* Cambridge University Press, Cambridge, UK.
3. Abramowiczy H, Abtt I, Adamczykh L, Adamusae M, Antonelli S, et al. Zeus Collaboration (2016) Limits on the Effective Quark Radius from Inclusive *ep* Scattering at HERA. *Phys Lett B* 757: 468–472.
4. Weinberg S (2008) *Cosmology.* Oxford University Press, New York.
5. Hawking SW, Ellis GFR (2023) *The Large Scale Structure of Space-Time: 50th Anniversary Edition.* Cambridge University Press, Cambridge, UK.
6. Profumo S (2017) *Introduction to Particle Dark Matter.* World Scientific Publishing, Singapore.
7. Fernández A (1985) Center-Manifold Extension of the Adiabatic-Elimination Method. *Phys Rev A* 32: 3070–3076.
8. Fernández A (1988) Phase-Ordering Dynamics for the Onset of a Center Manifold. *Phys Rev A* 38: 4256–4262.
9. Fernández A (1988) On Renormalization of Fluctuations at the Onset of a Centre Manifold. *J Phys A: Math Gen* 21: L607.

CHAPTER 5

Portal to the Dark Universe

"τὰ πάντα ῥεῖ καὶ οὐδὲν μένει"
"Everything flows and nothing stays."
HERACLITUS

This chapter establishes the topology of space-time, a feature not directly relevant to general relativity or quantum mechanics, the towering theoretical forays of the 20th century. As shown, the universe topology is essential to delineate the cosmic generation and depletion of dark energy and their bearing on the cosmological constant problem and on a dynamic equilibrium between dark and detectable matter. To that effect, the quantum fabric of space arising from vacuum entanglement is examined using the topological blueprint. Artificial intelligence, in the guise of a quantum gravity autoencoder with a physical embodiment (Chapter 3), reveals that a primeval wormhole underlies the topology of the universe and is sustained by a two-stroke cyclic cosmic engine fueled by vacuum energy. This engine maintains the baseline dark energy density observed experimentally, with dark energy getting replenished as vacuum energy fuels the universe runaway process associated with autocatalytic vacuum creation.

As shown, the cosmic engine sustains a portal to the dark universe in the form of a two-way primeval non-orientable wormhole (PNOW) whereby the "dark" right-handed fermions with spin 1/2 are in dynamic equilibrium with detectable matter. Dark fermions account for 18.72% of

DOI: 10.1201/9781003501534-6

dark matter. Since at present time "matter in all forms", including dark energy, is 26.7% dark matter and 5% detectable matter (Chapter 1), the dark fermion proportion at 18.72% of dark matter corresponds to 15.77% of all matter (dark + detectable) in the universe, which is exactly equal to the percentage of all matter represented by detectable matter.

We provide a perspective to interpret dark energy as the surplus in vacuum energy taken from the amount required to fuel the cosmic engine that maintains the PNOW. In contrast with the ever-decreasing concentration of dark and detectable matter that get progressively diluted in a growing vacuum, the concentration of the vacuum energy surplus is shown to be constant, as established by experiment. This vacuum energy surplus is identified as dark energy. The sustainability of the PNOW results from the coupling of vacuum energy expenditure with a dynamic equilibrium between dark fermions and visible matter, so that the overall concentration of dark energy is kept constant and so is the overall mass of dark matter. In this way, this chapter describes a cyclic cosmic engine that sustains a portal to the dark universe.

5.1 TOPOLOGY OF THE UNIVERSE AND THE SUSTAINABLE PRIMEVAL WORMHOLE

The global topology of the universe did not seem to be a matter of concern for Albert Einstein. We can speculate that this is so because his theory of gravitation, general relativity, is cast in terms of differential geometry and hence governed by differential equations. Since differential equations describe a local situation or setting, such as the behavior of curvature relative to mass distribution, the global topology of space-time plays no discernible role in general relativity. There is, however, a major constraint imposed by general relativity on the topology of space-time: The topology should remain invariant throughout the evolution of the universe. This is a difficult matter to grasp since time is part of the fabric of the object whose evolution we need to examine. Hence, as described in Chapters 2 and 4, we get into a sort of metaphysical conundrum whereby the object contains its own evolution as part of itself. To avoid semantic traps, we should state that by evolution of the universe we mean, by abuse of language, the changes in time of a generic constant-time cross section of space-time. There is no Russell-type paradox here, since space-time per se does not evolve, only space does, or rather the constant-time cross section of space-time actually evolves. Hence, the universe – actually meaning space – cannot change its topology throughout its evolution because that would mean that space-time presents an essential singularity in its fabric of the type forbidden by general relativity and that there is simply no well-defined topology that can be unambiguously attributed to space-time.

Quantum mechanics seems equally impervious to topological considerations. Quantum mechanics defines the fabric of space, as determined by vacuum entanglement, but topology is not factored into the theory in any obvious way, even if entanglement is essentially responsible for the nonlocality of quantum phenomenology. However, topology suddenly becomes extremely relevant as one tries to reconcile general relativity with quantum physics and identify the underpinnings of quantum gravity through the physical embodiment of a holographic autoencoder, as shown in Chapter 3.

If we assume topological invariance of space throughout the universe evolution starting at the big bang event, then the space of today cannot be infinite and flat, and therefore Euclidean, as Einstein would have it (he probably did not devote much attention to the matter). Provided we accept the big bang scenario, space today must be compact, not infinite, because it once was (i.e., 13.8 billion years ago), with certainty. But if space is compact, then it must be multiply connected, and that property bears pivotally on the solution of the cosmological constant problem, as we show subsequently.

In order to make further progress toward a complete characterization of space, we adopt a Parmenides-type of approach. This prompts us to formulate the question: Can space have a boundary? The answer is adamantly negative, since a boundary would imply an interface, not with the vacuum but with nothingness, and it is physically and metaphysically impossible to interface with nothingness: An interface presupposes two media, and nothingness is not a medium.

These considerations introduce several constraints on what the topology of the universe must be: It must be compact, locally flat, orientable, and lacking boundary. This leaves us with essentially one option, and it is a multiply connected 3-dimensional manifold: The 3-torus or Cartesian product $S^1 \times S^1 \times S^1$.

The fact that space is multiply connected introduces an extreme novelty and brings about hitherto unfathomable implications in the field of quantum gravity. To delineate such implications, we first remind ourselves of the way the 3-dimensional torus is constructed. It is essentially built by identifying opposite faces of a 3-dimensional cube (Figure 5.1). But if the aspect ratio χ_{ij} for every pair (i, j) of circular dimensions is $\chi_{ij} = \dfrac{r_i}{r_j} = 1$, as

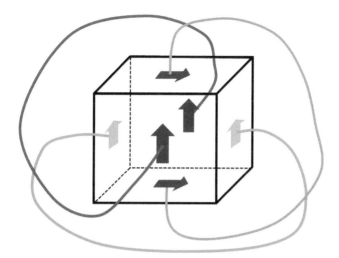

FIGURE 5.1 Topological identification of opposite faces of a 3-dimensional cube to render the 3-torus. The arrow bestows orientation to the face. Space is thus conceived as a compact manifold with no boundary, hence multiply connected.

mandated by the retention of symmetry upon coordinate permutation, then each 2-dimensional cross section of space becomes a horn torus and has a cusp-like curvature singularity. This cusp with its infinite spatial curvature is indicative of an entangled double black hole (BH), with interiors physically connected through an Einstein-Rosen (ER) bridge [1], as represented in Figure 5.2.

To represent the ER bridge in space-time requires higher powers of abstraction. We begin by representing the BH in space time, noting that

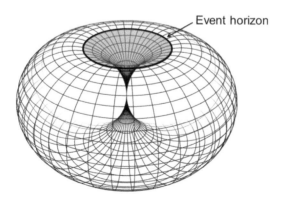

Event horizon

FIGURE 5.2 Toroidal 2-dimensional cross section of the universe along dimensions with aspect ratio 1. The cross section is a horn torus, hence generating a Schwarzschild double black hole with connected exteriors.

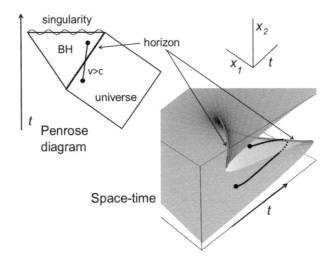

FIGURE 5.3   Representation of a black hole in a Penrose diagram (panel on the left, coordinate units scaled so that light travels at 45 degrees or speed 1), and in space-time (right panel).

the crossing of the event horizon entails going back in time, as indicated in Figure 5.3. Alternatively, to cross the event horizon requires a velocity higher than the speed of light, as suggested in the Penrose diagram displayed in Figure 5.3. Furthermore, the equivalent of a wormhole or Einstein-Rosen (ER) bridge between two black holes would be sustained between two exteriors (L and R) and two event horizons delineating interiors connected by the ER bridge. Thus, in space-time, the wormhole may be schematically represented as shown in Figure 5.4a,b.

As a solution to the Einstein equations, this entity is geometric, rather than physical (except that Einstein makes no distinction). In other words, there is no materiality in a double entangled BH, which becomes the realization of the solution to general relativity equations found by Karl Schwarzschild [1]. We name this solution the "primeval wormhole" (Figure 5.5) to distinguish it from the "material" types of shortcut BHs determined by extreme concentration of mass. The entanglement is obvious because the torus was created through identification, that is, perfect gluing of opposite faces (Figure 5.1), hence the physical ER connection in the horn torus begets an Einstein-Podolsky-Rosen (EPR) correlation of the opposite sides through the Maldacena-Susskind relation [2]. The "EPR=ER" relation is only valid for black holes, not for other entanglements, but the double cusp connection is indicative that we are precisely in that scenario.

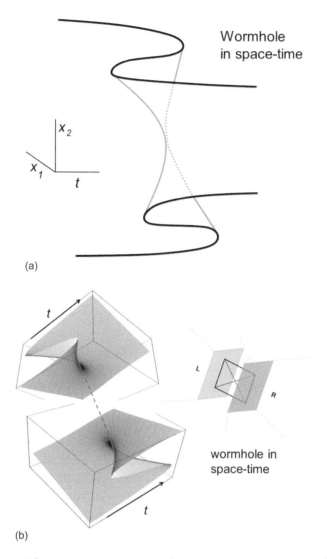

(a)

(b)

FIGURE 5.4 Schematic representation of a wormhole in space-time: (a) as two coupled folds of space time; (b) in relation to its Penrose diagram.

5.2 DECONSTRUCTING QUANTUM ENTANGLEMENT IN THE QUINTESSENTIAL TOPOLOGICAL BLUEPRINT OF SPACE-TIME

Entanglement is the most striking property of quantum physics because it enables us to assert the nonlocality of the quantum phenomena, in stark contrast with Newtonian physics or general relativity, which are local theories. The fabric of vacuum is clearly and solely a quantum attribute with

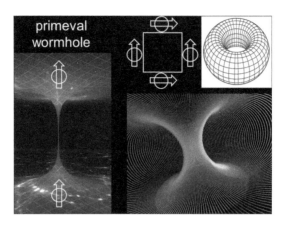

FIGURE 5.5 Entanglement of the primeval wormhole is postulated via the Maldacena-Susskind ansatz [1].

no clear relativistic counterpart and depends pivotally on entanglement. To understand the bearing of the fourth dormant spatial dimension on entanglement, we need to recapitulate the definition of entanglement in a guise that makes it amenable for quintessence deconstruction, without losing perspective of the central goal of this chapter: *Addressing the cosmological constant problem by incorporating the topological blueprint of the entangled universe.* In other words, we need to determine the bundle of compound states in space W that project onto a set of compound states in W/~ that contain the wave function for vacuum entanglement.

To fix notation, let us define a cell (spherical region) in space that is in two possible "ket states": $|1\rangle$ if the region contains a particle with associated de Broglie wavelength given by the cell dimension, and $|0\rangle$ if the cell is empty. Prior to measurement, which in quantum mechanics implies disambiguation, the region has an associated wave function $\frac{1}{\sqrt{2}}|0\rangle + \frac{1}{\sqrt{2}}|1\rangle$ Now, if we have two cells in space, the state of the compound system prior to measurement is tensorially represented in Dirac's simplified notation as

$$f_{00}|0\rangle|0\rangle + f_{01}|0\rangle|1\rangle + f_{10}|1\rangle|0\rangle + f_{11}|1\rangle|1\rangle, \tag{5.1}$$

with constants f_{ab}; a, $b = 0, 1$, satisfying the Born relation:

$$\sum_{a,b=0,1} f_{ab}^2 = 1. \tag{5.2}$$

The two cells are entangled if the following inequality is satisfied:

$$f_{00}f_{11} \neq f_{01}f_{10}. \tag{5.3}$$

So, we may adopt a measure of disentanglement defined as $\Delta \equiv |f_{00}f_{11} - f_{01}f_{10}|$.

For an arbitrary plane separating space in vacuum, two equidistant regions are maximally entangled, conforming a Bell pair with wave function (Figure 5.6):

$$\frac{1}{\sqrt{2}}|0\rangle|0\rangle + \frac{1}{\sqrt{2}}|1\rangle|1\rangle. \tag{5.4}$$

As we incorporate the fourth spatial "dormant" dimension, the Bell pair spans the following wave function in W:

$$\frac{1}{\sqrt{2}}\sin y_1|0\rangle|0\rangle + \frac{1}{\sqrt{2}}\cos y_1|0\rangle|1\rangle + \frac{1}{\sqrt{2}}\sin y_2|1\rangle|0\rangle + \frac{1}{\sqrt{2}}\cos y_2|1\rangle|1\rangle, \tag{5.5}$$

where the compound eigenstate coefficients are now functions of the dormant variable realized as y_1, y_2 for cells 1 and 2, respectively. The Bell pair

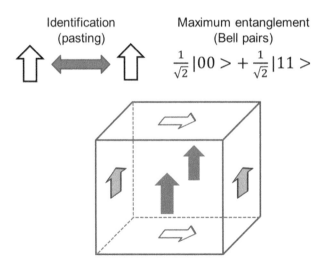

FIGURE 5.6 Topological identification of opposing faces in the 3-cube yields an entangled structure, whereby cells equidistant to the pasted surfaces are correlated via Bell pairs.

in quotient space W/~ is the projection obtained for $y_1 = \dfrac{\pi}{2}, y_2 = 0$. The entanglement bundle given by expression (5.5) becomes disentangled if and only if the following trigonometric relation holds:

$$\sin y_1 \cos y_2 = \sin y_2 \cos y_1 \tag{5.6}$$

which we may write as:

$$\frac{1}{2}\Big[\sin(y_1 + y_2) + \sin(y_1 - y_2)\Big] = \frac{1}{2}\Big[\sin(y_1 + y_2) + \sin(y_2 - y_1)\Big], \tag{5.7}$$

yielding:

$$\sin(y_1 - y_2) = 0 \Leftrightarrow y_1 = y_2. \tag{5.8}$$

Hence we have proven the following:

Lemma 5.1

The disentanglement set of states within the bundle in W associated with the Bell pair for vacuum entanglement in W/~ has measure zero. ▪

Theorem 5.1

The reversible work ΔG_δ required to disentangle the bundle in W to level $\Delta \leq \delta$ is

$$\Delta G_\delta = -T\Delta S_\delta = -k_B T \ln\left(\frac{\mu \mathcal{E}_\delta}{\mu D_\delta}\right) \tag{5.9}$$

where $\mathcal{E}_\delta$ is the region associated with disentanglement level $\Delta \leq \delta$ (Figure 5.7), μ denotes measure, and $\mu D_\delta = 4\pi^2 - \mu \mathcal{E}_\delta$.

Proof: The reversible work or free energy increment required to disentangle the bundle to level $\Delta \leq \delta$ is computed as entropy loss ($\Delta G = - T\Delta S_\delta$), hence determined by the quotient of the multiplicities of disentangled over entangled states in the bundle. Full disentanglement ($\Delta = 0$) requires infinite work, as per Lemma 5.1. ▪

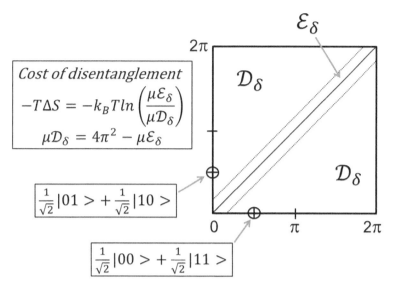

FIGURE 5.7 Region and dark energy cost of disentanglement to level δ for the bundle in W that subsumes the Bell pair of maximal spatial entanglement in W/~.

5.3 AI PROBES THE EQUIVALENCE BETWEEN WORMHOLE AND QUANTUM ENTANGLEMENT

The elucidation of the nature of dark matter in Chapter 4 highlights the power of topology-based metamodels in leveraging autoencoder technology for model discovery. The main assertion in Chapter 3 is that a neural network with emergent gravity admits an autoencoder with emergent quantum behavior and hidden variables that can be modeled with relativistic strings. Thus, a quantum metamodel of gravity is in principle possible. This finding paves the way to tackle the next conundrum involving the conjectured equivalence between relativistic wormholes and quantum entanglement.

Spacetime locality is a basic tenet of modern physics. The term locality refers to the impossibility of sending signals at speeds higher than the speed of light, an idea that is constantly challenged both by quantum mechanics and general relativity. Thus, quantum mechanics gave rise to the Einstein-Podolsky-Rosen (EPR) correlations also termed "entanglements," while general relativity allows for solutions to the equations of motion that connect distant regions through short-circuiting "wormholes" also known as Einstein-Rosen (ER) bridges [1]. Physicists Maldacena and Susskind [2] have conjectured that these two concepts may be connected by a duality that becomes in effect equivalence, akin to the similar duality found in the physical underpinnings of quantum gravity. They have argued persuasively that the ER bridge between two black holes may be actually created by EPR correlations between microstates of the two black holes and labelled the conjecture "ER = EPR relation". In their analysis, the ER bridge is a type of EPR correlation in which the correlated quantum systems are in a specific entangled state that admits a weakly coupled Einstein gravity description. This situation is illustrated by a black hole pair creation in a magnetic field, and it is tempting to think that any EPR–correlated system, even a simple singlet state of two spins, is connected by some sort of ER bridge.

The neural network model of the universe endowed with the quantum gravity autoencoder described in Chapter 3 seems an ideal system to validate (or disprove) the EPR=ER relation. This is because the emergent quantum behavior of the autoencoder pivots on hidden variables that are interacting through relativistic strings, and the very existence of the hidden variables is known to resolve the EPR paradox. In the quantum gravity autoencoder, the trainable variables conforming the **q**-vector exhibit a quantum mechanical behavior in an equilibrium regime where the network state variables that constitute the **x**-vector have been thermalized. A learning process involves L separate sets of training **x**-vectors with expected values $\bar{x}^l, l = 1, 2, \ldots, L$, and the expectation vectors together with the expectation state vector $\bar{x}^0$, representing network evolution to zeroth order in the linear approximation to node activation, are regarded as the hidden variables in the emergent quantum behavior of the trainable **q**-states. On the other hand, the nonequilibrium dynamics of the hidden variables becomes relevant on timescales much smaller than their thermalization time. This nonequilibrium dynamics is determined by the strength of the weak interactions between vector pairs $\bar{x}^\nu$, $\bar{x}^\xi$, $\nu, \xi = 0, 1, \ldots, L$ quantified by the tensor $g_{\nu\xi}^i$, where the dummy index i labels each neuron in the system. By weak interactions we mean that the generic vectors $\bar{x}^\nu$, $\bar{x}^\xi$ are not interacting directly but through the **q**-vector that they themselves contribute to train. These dynamics can be cast in terms of relativistic strings in an emergent space-time, as indicated in Chapter 3.

This AI framework with emergent physics seems ideal to validate the EPR=ER relation. If two parts of the **q**-space are entangled in the quantum autoencoder, they should be bridged through a wormhole in the full network with emergent gravity, as described in Figure 5.8. The presumed proportionality between the gravitational action and the quantum action

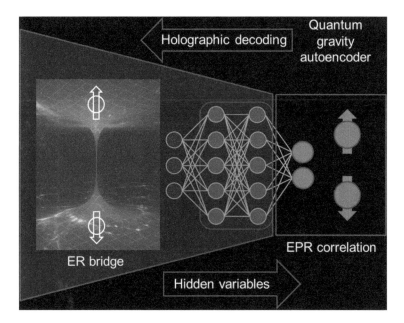

FIGURE 5.8 Schematics of quantum gravity autoencoder (QGAE, Chapter 3) deployed to validate the "EPR=ER relation."

may be a good starting point for this project aimed at testing the power of AI as purveyor of the physical underpinnings for the most perplexing problems in modern cosmology.

The narrative in this book takes an unexpected turn in Chapter 3, where it is shown that a connected array of neurons (NN) may be treated not only as an information-processing machine but also as a statistical mechanical object, capable of exhibiting emergent physical behavior. Thus, a duality between emergent gravity and quantum mechanics is established through an autoencoder that thermalizes hidden degrees of freedom arrayed on the NN state vector $\mathbf{x}$. In this way, the autoencoder exhibits emergent quantum behavior while the nonequilibrium dynamics of the hidden variables spans an emergent space-time endowed with gravity (Chapter 3). This physical duality of the learning machine cast in terms of statistical mechanics should enable topological innovation on the emergent space-time through quantum entanglement at the autoencoder level.

To illustrate this point, consider two completely separated replicas of a quantum gravity autoencoder (QGAE), labelled left (L) and right (R) with identical collection of training sets $\left\{ \mathcal{S}^{(\mu)} \right\}_{\mu}$ comprised of $\mathbf{x}$-vectors and generating the sets of $\mathbf{q}$-vectors $Q_L = \left\{ Argmin\, J_<^{(\mu,L)} \right\}_{\mu}, Q_R = \left\{ Argmin\, J_<^{(\mu,R)} \right\}_{\mu}$, in

QGAE(L) and QGAE(R), respectively. Notice that the stochastic nature of the training process implies that $Argmin\ J_<^{(\mu,L)}$ is not necessarily equal to $Argmin\ J_<^{(\mu,R)}$, but surely the q-vector pairs $Argmin\ J_<^{(\mu,L)}$, $Argmin\ J_<^{(\mu,R)}$ are entangled as they have a common origin. Hence if the NNs are architecturally configured so that there is a black hole in Q_L and therefore in Q_R, then both black holes are necessarily entangled. This entanglement begets connectivity in the respective emergent space-times associated with QGAE(L) and QGAE(R), which would be topologically related to the Penrose diagram shown in Figure 5.9.

Thus, the equivalent of a wormhole or ER bridge between the black holes would be sustained between the two learning machines L and R as a result of entanglement at the QGAE level. This "double black hole" would have two exteriors (L and R) and two event horizons meeting at the counterpart of the ER bridge shown in the Penrose diagram (Figure 5.9). However, we expect this sort of connectivity to be topologically different in the case of entangled NNs. This is so because the emergent space-time constructed upon the q-space of a QGAE is essentially different from the Minkowski space where the particular solution to Einstein's equations was obtained and topologically described by the Penrose diagram.

Describing the spacetime topology of entangled NNs may prove rewarding, as it is likely to herald a new breed of quantum computation that we may provisionally term "quantum gravity computation."

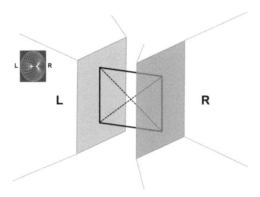

FIGURE 5.9 Penrose diagram counterpart for the connection topology in emergent space-time arising from the entanglement of two black holes within quantum gravity autoencoders (L and R). The respective event horizons are marked by dashed-line upper and lower triangles in a conformal 2-dimensional spacetime.

5.4 PRIMEVAL WORMHOLE AS A TWO-STROKE ENTANGLE/DISENTANGLE ENGINE FUELED BY DARK MATTER: A PROVISIONAL SOLUTION TO THE COSMOLOGICAL CONSTANT PROBLEM

Theorem 5.1 enables further understanding of the entanglement of the primeval wormhole. To further investigate this matter, we implement our AI tool with an architecture consisting of an annular arrangement of two replicas of a QGAE (Figure 5.10). The emerging results reveal that a massive expenditure in dark energy is required for an interim destruction of the primeval wormhole by changing the aspect ratio $\chi = 1 \rightarrow \chi > 1$, resulting in the conversion of a horn torus into a spindle torus. As per the Maldacena-Susskind ansatz [1], this change in dimensional aspect ratio must correlate with a change $\Delta = 0 \rightarrow \Delta \leq \delta$ in the level of wormhole disentanglement, with $\delta = function\ of\ (\chi - 1)$ (Figure 5.11).

Thus, Figure 5.11 displays the dark energy expenditure in terms of the level of disentanglement ($\Delta \leq \delta$), and equivalently, in terms of the change in aspect ratio, quantified as $(\chi - 1) > 0$, which indicates the dismantling of the ER bridge. In other words, the dark energy expenditure for disentanglement of the primeval wormhole generates or rather enables an increase in one circular dimension, say, r_2, relative to the other circular dimension with radius r_1. We are assuming without loss of generality that

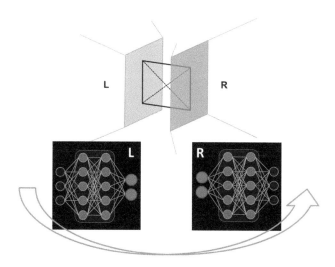

FIGURE 5.10   Annular arrangement of two replicas of the holographic (quantum gravity) autoencoder as required to investigate the sustainability and dynamics of the primeval wormhole arising in the 3-toroidal space.

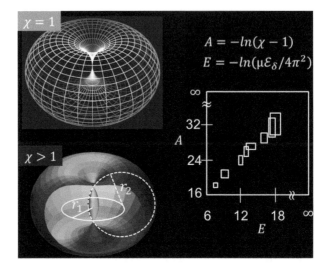

FIGURE 5.11 The two-stroke engine sustaining the primeval wormhole and fueled by dark energy. Correlation between disentanglement level (δ) and change in aspect ratio χ due to dimensional expansion.

the aspect ratio is given as $\chi = \dfrac{r_2}{r_1}$. The universe expansion along one dimension is enabled and materializes because the primeval wormhole gives back the dark energy invested in its disentanglement, while additional dark energy is generated as the volume of the vacuum increases with the expansion. This surplus in dark energy enables its reinvestment in the next cycle of entanglement-disentanglement that has a higher cost in terms of reversible work.

As we can see, the chain of events:

[δ-level disentanglement] → [vacuum expansion to restore $\chi \approx 1$] → [δ-level disentanglement] describes *the primeval wormhole as a two-stroke engine fueled by dark energy* (Figure 5.12), whereby the measured density of dark energy (ρ_Λ) is given as:

$$\rho_\Lambda = \rho_e\left(\delta\right) + \rho_{\Delta V}\left(\delta\right) - \rho_{e'}\left(\delta\right), \tag{5.10}$$

where $\rho_e(\delta)$ is the density of dark energy released upon restoration of the entanglement from a δ-level disentanglement of the primeval wormhole, $\rho_{\Delta V}(\delta)$ is the surplus in dark energy per unit volume representing the extra vacuum energy associated with the vacuum dimensional expansion (ΔV)

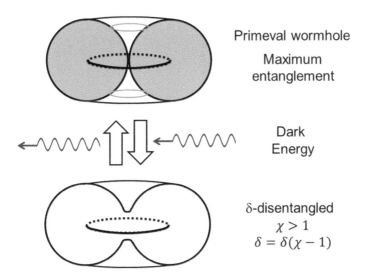

Primeval wormhole

Maximum
entanglement

Dark
Energy

δ-disentangled
$\chi > 1$
$\delta = \delta(\chi - 1)$

FIGURE 5.12   Representation of one cycle of the cosmic engine that sustains the primeval wormhole.

that restores the entanglement ($\chi \approx 1$), and $\rho_{e'}(\delta)$ is the next dark energy investment per unit volume required to disentangle the primeval wormhole to δ-level. Since $\rho_\Lambda \approx 5 \times 10^{-10} J/m^3$, we obtain $\delta = \delta_\Lambda = 34.01 \times 10^{-18}$.

Space in the 3-torus universe grows one dimension at a time, through cycles associated with enormous bursts of dark matter with restitution of the primeval wormhole entanglement followed by ever-increasing expenditures of dark energy required to disentangle the reconstituted primeval wormhole. The amount of dark energy that is effectively not spent in the cycle may be computed as the difference between two consecutive bursts.

Thus, the cosmological constant problem, often described as defined by the enormous (120 orders of magnitude) surplus of dark energy arising from vacuum fluctuations, is more subtle than what it appears to be in the standard formulation [3]: *Dark energy is spent at an ever-increasing cost and replenished cyclically by a pulsating universe that tears and regenerates the entanglement fabric of its primeval wormhole through the alternating expansion of its vacuum compact dimensions.* This endless sequence of cycles of destruction and creation in the evolution of the universe shares an astonishing parallel with the ancient cosmogonies stemming from metaphysical inquiries in the valley of the Indus. The infinite toils of Shiva, the destructive deity that as "Nataraja" dances out the pulse of the

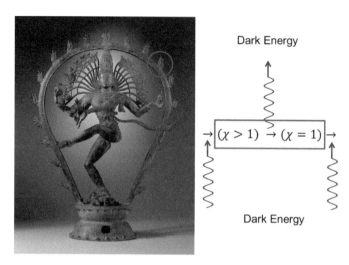

FIGURE 5.13   Shiva as "Nataraja", the lord of the dance, the destructive Hindu deity that also dictates the rhythm of the universe. The tempo of the pulsating universe is marked by Shiva's hourglass-shaped drum. As Nataraja, Shiva dances out the sequence of cycles of successive destructions followed by reconstructions of the universe that determine its evolution. Each cycle in the Nataraja's dance is marked by a flame (circle), akin to the burst of dark energy released with the reconstruction of the entanglement fabric of the primeval wormhole. This cosmic event is followed by the reinvestment of the dark energy in the next cycle that begins with the destruction of the entanglement of the primeval wormhole. Credit: Los Angeles County Museum of Art. Image of the Hindu deity "Shiva Nataraja" obtained from the public domain (https://en.wikipedia.org/wiki/File: Shiva_as_the_Lord_of_Dance_LACMA_edit.jpg).

universe, is particularly suggestive of the rhythm conferred to the expansion of the universe through the cycles of destruction and creation of the entanglement fabric of its primeval wormhole. Shiva's dance is strikingly suggestive of the topology-sustaining two-stroke engine fueled by dark energy and controlling the vacuum expansion in the universe runaway (Figure 5.13).

5.5 A HOLISTIC QUINTESSENTIALLY ENTANGLED UNIVERSE ADMITS NO OBSERVER

The state of a cell in the entangled fabric of quantum vacuum is of course ambiguous, as it befits the standard interpretation of quantum mechanics, whereby disambiguation requires an act of observation. The structure of vacuum is upheld through the correlation pattern defined by the compounded ambiguous state: $\frac{1}{\sqrt{2}}|0\rangle|0\rangle + \frac{1}{\sqrt{2}}|1\rangle|1\rangle$, a Bell pair indicative of maximal entanglement. The collapse of the wavefunction for one of the cells, say cell 1, would yield ket state $|0\rangle$ or $|1\rangle$, depending on whether the observation registers respectively the absence or presence of a particle with associated de Broglie wave equal the dimension of the cell. This observation instantly triggers the collapse of the wave function for cell 2. Thus, the state of the system is now either $|0\rangle|0\rangle$ or $|1\rangle|1\rangle$, depending on the measurement that took place for cell 1. This is a completely standard reasoning befitting the so-called Copenhagen interpretation and holds no surprise. It fits exactly with the dictum by John A. Wheeler: "No phenomenon is a physical phenomenon until it is an observed phenomenon."

However, the act of observation has been placed and can only be placed in the quotient space W/~, a latent manifold which is a projection of the quintessential space W. But, as we examine the quintessential entanglement bundle (Eq. 5.5) corresponding to the Bell pair (Eq. 5.4) in quotient space, we notice that there is no component of the bundle that projects on either $|0\rangle|0\rangle$ or $|1\rangle|1\rangle$. In other words, there is no value pair for the dormant variables y_1, y_2 for cells 1 and 2 that would yield either $|0\rangle|0\rangle$ or $|1\rangle|1\rangle$ as projection onto W/~.

This indicates that entanglement is never lost in W as a result of observation. We may formalize this observation as follows:

Theorem 5.2

Space W does not admit quantum observation, hence it should be considered quintessentially entangled. ■

The quintessential entanglement is again strongly suggested by the ancient metaphysical cogitations from Hinduism that posit the holistic nature of the universe as a major premise in its cosmogony. In this cosmogony developed over four millennia ago by the civilization of the Indus valley, all things and beings stem from a primeval unity and are unified by their blending into *Brahman*, the ultimate holistic entity, designated by the ōm symbol (Figure 5.14). Strikingly, it is precisely the ōm symbol that is believed to be the sign produced by the figure of Shiva Nataraja in its choreography of the pulsating universe marked by cycles of destruction and reconstruction (Figure 5.13).

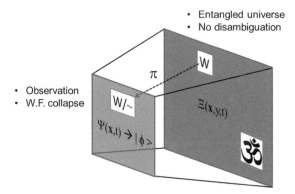

FIGURE 5.14 The quintessential universe W admits no observer, hence all quantum phenomena remain potential and unrealized. The quintessential wave function does not collapse with observation, in contrast with its projection onto the latent space W/~. The holistic nature of W is thus apparent and may be equated with *Brahman*, whose sign "ōm" is featured in the figure.

5.6 PRESERVATION OF THE TOPOLOGY AND QUANTUM FABRIC OF SPACE-TIME IN A MULTIVERSE SCENARIO

AI has addressed the quantum gravity problem in more than one way, and one alternative is constructing a "physical learning machine" or a "learning machine with a physical embodiment," with stochastic connectivity weights and hidden variables representing the random states of the nodes. This system may be endowed with emergent gravity and emergent quantum behavior (Chapter 3). In this way, AI represents the universe as a holographic autoencoder, where the holographic map $h : \partial W \to W$ is one of the many possible inverses of the canonical projection $\pi : W \to \partial W$ that maps the quintessential space W onto its "quantum border" ∂W. The emergent quantum behavior arises in the equilibrium regime for node state (hidden) variables upon which relativistic strings are embroidered. Conversely, the network is endowed with an emergent gravity in the nonequilibrium regime for the hidden variables. The results illustrate the possibility of establishing the duality quantum mechanics/ general relativity in neural networks regarded as physical systems governed by the laws of statistical mechanics (Chapter 3). These networks are endowed with two different thermodynamic regimes, where the geometric dilution parameter $v = -\log \cos\alpha$ is the proxy for time and the advent of the lightest dark matter occurs at $\sim 10^{-27}s$ counting from the birth of the universe.

The coupling of two replicas of the QGAE in an annular conformation (Figure 5.10) was considered a suitable arrangement in order to reproduce the primeval wormhole arising from the construction of space as the 3-torus. But this AI setting does not tell us whether $\delta = \delta_\Lambda = 34.01 \times 10^{-18}$ represents a universal constant that actually materializes in the universe. If it does not, then the dark matter density of the universe would not be constant, but we know that it is, at least approximately. There is in principle no *a priori* reason for the level of disentanglement of the primeval wormhole not to vary with each cycle. If it happens to allow for a surplus or shortfall in dark energy, then an additional cosmic mechanism would need to be invoked, representing a sort of "cosmic valve" that needs to become operative to maintain a constant density of dark matter energy, as mandated by the cosmological constant. This cosmic valve mechanism turns out to be cosmic reproduction.

As described in Chapter 3, a birth channel for cosmic reproduction can be set up within this scheme by constructing a second holographic auto-encoder $h' : \partial W' \to W'$ that serves as antenna, capable of capturing the tunneling $\partial W \to \partial W'$ of a dark energy burst in the universe $(\partial W, W)$ in the form of a quantum vacuum fluctuation. Quantum tunneling implies the existence of a spatial interface, and the only spatial dimension of the universe that offers a boundary is the dormant dimension, hence the tunneling would have to be "quintessential," that is, operative through the dormant dimension. The tunneled fluctuation is amplified within the second holographic autoencoder to give birth to a progeny universe $(\partial W', W')$. By $\sim 10^{-27}s$ after its birth, the baby universe will begin storing the tunneled dark energy in a material embodiment, with the formation of dark matter in the form of ur-Higgs particles. In this way, AI constructs a cosmic reproductive machine that converts dark energy into dark matter (Figure 5.15). Notably, this machine harnesses dark energy as the cosmic "birth inducer" and generates dark matter in a multiverse scenario as the canonical (surjective but not injective) projection $\pi' : W' \to \partial W'$ admits *a priori* a multiplicity of holographic maps $h' : \partial W' \to W'$ all satisfying the equation $h' \circ \pi' = id_{W'}$.

To leverage this multiverse scenario, an advanced AI–based cosmic technology for universe reproduction is implemented (Chapter 3). The technology harnesses dark energy tunneling as a nucleating quantum fluctuation that spans the progeny universe. The cosmic reproduction machine requires a second holographic autoencoder serving as antenna or $\partial W \to \partial W'$ receiver. An energy surplus from the ever-increasing vacuum

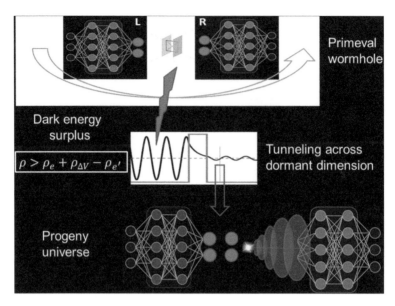

FIGURE 5.15   AI–based cosmic reproductive machine that converts dark energy into dark matter in a multiverse matrix. A birth canal for cosmic reproduction is set up by coupling the universe ($\Omega = \partial W$, W) to a second holographic autoencoder acting as receiver and capable of capturing the tunneled dark energy originated in the progenitor universe. The tunneled quantum fluctuation is amplified through the receiver autoencoder to give birth to the progeny universe ($\Omega' = \partial W'$, W') that stores the tunneled dark energy as dark matter in the form of stationary waves along the quintessential circular coordinate of quark-size attometer dimension.

quantum energy that is not invested in restoring the primeval wormhole entanglement is meant to be indicative of a poor understanding of dark energy in our universe. This surplus may be reconciled within the multiverse scenario put together by a technologically advanced civilization capable of cosmic investment in dark energy for reproductive purposes, as schematically depicted in Figure 5.15.

On the other hand, the cosmic valve may work in the opposite direction as required in the event there is a shortfall in dark energy to restore the entanglement of the primeval wormhole while maintaining the dark energy density at its constant value $\rho_\Lambda \approx 5 \times 10^{-10} J/m^3$. In this case the conversion of dark matter into dark energy and tunneling of the latter through the interface of the dormant dimension would be carried out by the lender, which now is the progeny universe.

The vacuum catastrophe or cosmological constant problem (Chapter 1) alluded to earlier is arguably the most embarrassing discrepancy in all of physics. It has become a veritable "tag for demolition" of the whole edifice of physics. A naïve computation of the vacuum quantum energy density yields values estimated to be 120 orders of magnitude larger than the expected contribution to the cosmological constant based on experimental observation [3]. AI provides a solution to this problem, shifting the task to a mere calculation of the rate of generation of progeny universes arising from dark energy tunneling. Thus, the tunneling of vacuum fluctuations is given a material embodiment as dark-matter ur-Higgs particles in the embryonic universe. These particles are stored as stable wave excitations along the quintessential quark-scale dimension and endowed with mass 246GeV/c^2, the vacuum expectation value of the Higgs boson.

We have described AI's multiverse solution to the cosmological constant problem. The solution is based on a "multiverse matrix," a pivotal scenario consisting of coupled holographic autoencoders capable of converting dark energy into dark matter as the tunneling of dark energy is steered through a cosmic birth canal that exploits the dormant dimension. The multiverse generator acts as a cosmic valve to maintain a constant dark energy density in the case of surplus or shortfall in the fueling of the two-stroke engine that holds together the topology of the universe and its fabric of quantum entanglement.

5.7 CHIRALITY IN THE DARK UNIVERSE

The invisibility of dark matter is directly attributable to a lack of communication with the weak force. This statement implies that a proportion of the negative-energy fermionic sea, known as the Dirac sea (Chapter 1), must be somehow rendered impervious to communication of the weak force, but how? Identifying the actual purveyor to the dark universe requires a careful discussion of "chirality," an essential concept in particle physics.

It is obviously tempting to identify the Dirac sea of negative-energy fermions, or the closely related vacuum energy in quantum field theory, with a portion of the dark universe. Taken at face value, this assertion is plainly incorrect. A revision and fine-tuning of the previous material presented in this chapter becomes mandatory as we confront the fact that the weak interaction that enables fermions to escape the Dirac sea (with concurrent creation of particle-antiparticle pairs) is *chiral*, a concept now to be discussed.

Chirality is a property of particles that has not entered so far into the discussion put forth in this book. The decisive experiment by Chien-Shiung Wu, asserting the lack of parity conservation (left/right handedness equivalence) for the weak interaction [4–6], makes this discussion imperative. It was this experiment that prompted the unification of the weak force with electromagnetism, a merging now named electroweak force.

Handedness may be defined by the sign of the projection of spin axis of rotation along the direction of motion (left = negative, right = positive), and it is a constant of motion but not a Lorentz-invariant except in the case of massless particles, where handedness is equal to chirality. For massive particles, handedness is obviously not Lorentz-invariant, since a frame of reference may move faster than the particle, reversing the particle's direction of motion relative to the observer. Thus, handedness and chirality cannot be equated for massive particles, with the latter being Lorentz-invariant but not a constant of motion, *au contraire* to handedness.

Spin ½ fermions, and in particular electrons, generated in vacuum fluctuations as particle-antiparticle pairs, as described schematically by the Feynman diagram in Figure 5.16, are left-handed, in accord with the chirality of the weak force. Likewise, if the universe arose from a vacuum fluctuation as asserted in Chapter 4, we would expect that the Dirac sea would be initially composed of left-handed fermions, hence not part of the dark universe *per se*. So, the identification of vacuum energy with dark energy as put forth in this chapter must be upheld with a major caveat.

The autoencoder endowed with its physical realization (Chapter 3) is suited to materialize a holographic universe, since the information embossed in space-time is encoded in the latent manifold endowed with a fabric of quantum entanglement. However, no constraint arising from the existence of invisible matter (or its energetic equivalent) has been imposed on the encoding of the atlas – that is, the covering with locally tangent spaces – of the latent manifold. In other words, the existence of a dark universe was not factored into the construction of the latent manifold on the same footing as the constraints (compactness and connectedness) that

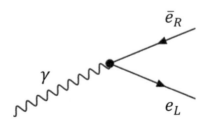

FIGURE 5.16   Feynman diagram for the vacuum creation of the electron-positron pair with the respective particle chiralities.

determined its topology, with its primeval orientable wormhole (POW). As now shown, those constraints pertain to the *orientability* of the manifold, a property that should be topologically on equal footing as compactness and connectedness.

To warrant their invisibility, fermions in the dark universe must be endowed with right-handed chirality. On the other hand, orientability is the only topological feature that can be tuned at this point in the discussion to accommodate this possibility, as argued in the subsequent sections. We are compelled to admit that the previous development requires revision, as the latent manifold may need to be reoriented if this possibility impinges on chirality, as it indeed does. More specifically, we are seeking a portal to the dark universe enshrined in the topology of the universe.

5.8 PRIMEVAL NONORIENTABLE WORMHOLE AS PORTAL TO THE DARK UNIVERSE

In defining the topology of the universe, we have established that space-time as a manifold needs to be compact and multiply connected since it is constrained to have no boundary. However, we have not yet dealt with orientability, a topological latitude that, as it turns out, plays a pivotal role in determining the portal to the dark universe.

As delineated in the previous sections, the primeval wormhole is sculpted by the topological constraints to which space-time is subject. On the other hand, the setting of "universe as hologram" fits into an AI system in the guise of an autoencoder, precisely the system that captures the essential physics with dimensionality reduction. This autoencoder has been subject to constraints that defined the encoding of the tangent bundle of its latent manifold. However, the encoding of the tangent bundle is itself subject to constraints that have not been hitherto considered and result from the tuning of orientability.

To better understand the problem, we need to identify the portal for the generation of the "dark" dextrogyre (right-handed) spin ½ fermions. These dark fermions are actually specular versions of those generated in pair creation through vacuum fluctuation (Figure 1.19). The concept of mirror matter was introduced by Lee and Yang with the idea of parity (P) violation in weak interactions [6]. The mirror "hidden" matter sector of quantum field theory restores the equivalence in the universe under mirror reflection. As it has been established, the Standard Model Lagrangian contains only the left-handed components of the Dirac wave functions: $\psi_L = P_L\psi$, where the left-handed and right-handed projectors are made up of products of Dirac matrices (Figure 1.18):

$$P_L = \frac{1}{2}\left(\mathbb{I} - \gamma^5\right); P_R = \frac{1}{2}\left(\mathbb{I} + \gamma^5\right); \gamma^5 = i\gamma^0\gamma^1\gamma^2\gamma^3. \tag{5.11}$$

Within the Standard Model, left and right chirality components are treated differently by the weak interaction. For example, the weak interaction could rotate a left-handed electron into a left-handed neutrino with emission of a boson W⁻, but could not do so with right-handed counterparts. Therefore, we need to tune the topological latitude of the primeval wormhole so that, upon crossing, it should enable the transformation:

$$P_L = \frac{1}{2}\left(\mathbb{I} - \gamma^5\right) \rightarrow P_R = \frac{1}{2}\left(\mathbb{I} + \gamma^5\right); \mathbb{I} = identity\ matrix. \tag{5.12}$$

As it turns out, a primeval non-orientable wormhole (PNOW) would enable such transformation [7] and therefore may serve as a portal to the dark universe made up of right-handed Dirac fermions. This implies that the entanglement pattern arising from boundary identification in the universe should – in all likelihood – be non-orientable, as described in Figures 5.17a,b. This quotient space may be represented as the Cartesian product of a 2-torus and a Klein bottle (Figure 5.18). It should be emphasized that the hole in the non-orientable factor-manifold does not exist in four or more dimensions: It is purely artifactual since it results from the inherent limitations of the 3-dimensional representation.

To construct a 2-dimensional cross section of the PNOW as purveyor of dark fermions, we need to join the exteriors of event horizons for the two connected black holes but replace an orientable (Figure 5.19) for a non-orientable surface obtained by cutting and pasting a Moebius strip (Figures 5.20, 5.21a). The actual AI–system (coupled autoencoders) realizing the non-orientable wormhole is described in Figure 5.21b, whereby the tangent bundles for the respective latent manifolds are encoded with identification following the reflection symmetry: Two tangent spaces that transform into each other through the symmetry operation are identified in the Moebius-type annular arrangement of quantum gravity (holographic) autoencoders, as shown schematically in Figure 5.21b. The entire encoded atlases are thus glued steered by the inversion operation.

One would be tempted to assume that the PNOW could change the chirality of the negative-energy fermions in the Dirac sea upon crossing

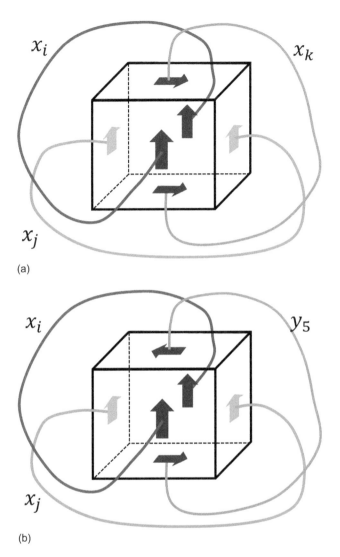

(a)

(b)

FIGURE 5.17 Topological identification of opposite faces of a 3-dimensional cube to render a 3-torus **(a)** or a non-orientable 3-dimensional manifold **(b)**, depending on the space-time coordinates considered (x denotes any of the four canonical, while y_5 is the fifth compact coordinate). The arrow bestows orientation to the face. In topological terms, the non-orientable manifold **(b)** is a Cartesian product of a circle and a Klein bottle.

(Figure 5.22), but such picture would not be altogether compatible with quantum field theory, which supersedes the Dirac ansatz in favor of particle-pair creation through vacuum fluctuations, as discussed in Chapter 1 (Figure 1.19). Thus the correct picture of the PNOW as portal for the dark

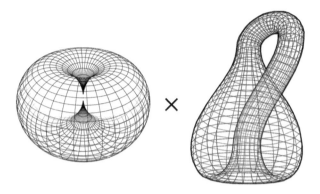

FIGURE 5.18 Four-dimensional non-orientable spatial cross section of the quintessential space-time represented as the Cartesian product of a 2-torus and a Klein bottle.

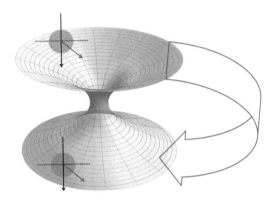

FIGURE 5.19 Topological representation of the primeval orientable wormhole (POW) that preserves particle chirality upon crossing (spin axis as red vector, momentum vector in black).

universe is provided in Figure 5.23. The right-handed "dark" fermion generated by the PNOW is precluded from contributing to the replenishing of the Dirac sea because of its lack of interaction with gauge bosons that communicate the weak force and would enable transition to the lower energy levels. Nevertheless, crossing cannot be only one way, since that would imply a constant production of dark matter, which would be at odds with what is established in that regard: *The total amount of dark matter remains constant*, in contrast with the total amount of dark energy that grows constantly to maintain constant concentration upon vacuum growth. This implies that transitions yielding dark fermions are reversible in the sense of

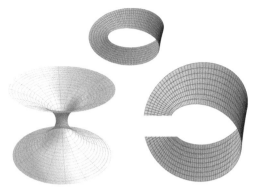

FIGURE 5.20   Elements for the construction of a 2-dimensional cross section of a non-orientable wormhole by cutting and pasting a Moebius strip.

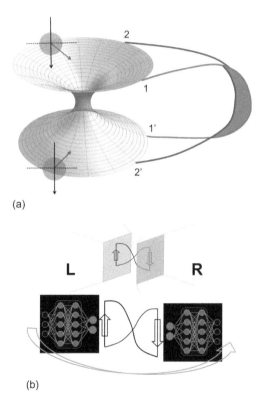

FIGURE 5.21   Topological representation of a primeval non-orientable wormhole (PNOW). (a) 2-dimensional rendering of the manifold displaying through surgical attachment of a Moebius strip with inversion of particle chirality upon crossing. (b) Annular assemblage of two holographic autoencoders with outer layers connected in opposite orientations.

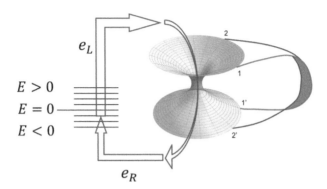

FIGURE 5.22 PNOW as direct supplier of dextrogyre "dark" spin-1/2 fermions to the Dirac sea of negative-energy fermions.

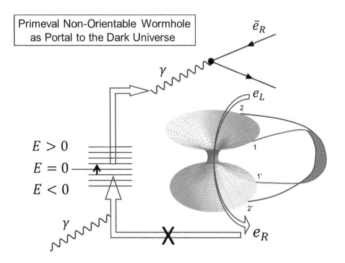

FIGURE 5.23 PNOW as cosmic engine that sustains the portal to the dark universe in the quantum field theoretical vacuum. The "negative energy flow" drawn from the Dirac sea maintains the PNOW, which is thus fueled by dark energy in the quantum field interpretation of vacuum energy. The coupling between fueling dark energy and production of dark fermions materializes in the operation of the cosmic engine.

being in dynamic equilibrium with the transitions of dark fermions into visible levogyre or detectable fermions via the crossing of the PNOW, as described in Figure 5.24.

Through the process described in Figure 5.24, the PNOW can replenish the depleted Dirac sea, which only gets repopulated from detectable "levogyre" fermions upon photon emission.

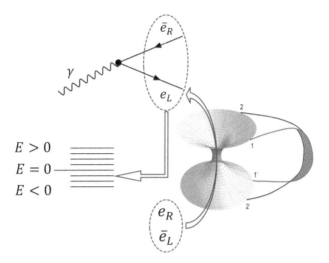

FIGURE 5.24 Inverse transformation of dark spin-1/2 fermions into visible fermions upon crossing the PNOW. The transformations of matter described in Figures 5.23 and 5.24 are in dynamic equilibrium.

5.9 CONSTRUCTING A PORTAL TO THE DARK UNIVERSE IN A QUANTUM COMPUTER

The previous sections have described the sustainability of the primeval wormhole – and, thereby, of the topology of the universe – through constant expenditure of dark energy in an expanding vacuum generated autocatalytically. Thus, an open traversable wormhole fitting the universe's horn-torus topology was shown to be sustained by a cosmic engine fueled by dark energy. The engine was shown to partially disentangle the exteriors of the primeval wormhole ($\chi = 1 \rightarrow \chi > 1$), enabling autocatalytic universe expansion, one dimension at a time. In turn, this process restores entanglement and replenishes dark energy, completing the cycle of the cosmic two-stroke engine. Since dark energy is the ever increasing "vacuum energy surplus" after fueling the cosmic engine through its successive cycles, this implies that a shockwave of "negative energy" (i.e., lower than the expected vacuum energy) must traverse the wormhole at regular intervals to keep it open, hence enabling information passage. This negative energy traffic across the PNOW is described in Figure 5.23.

This prompts us to address the possibility of creating the primeval wormhole in a quantum computer, taking advantage of the possibility of manipulating quantum entanglement of qubits which are embodied in the space-time wormhole in accord with the Maldacena-Susskind conjecture [1].

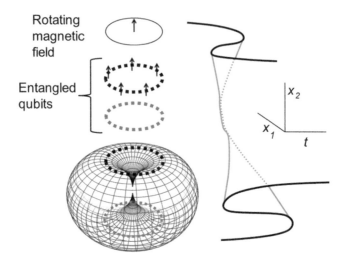

FIGURE 5.25 Dark energy as enabler and sustainer of the primeval wormhole recreated in a quantum computer as a quasi-continuum "pulse of dark energy deficiency" sustained between the two wormhole exteriors and traversing its interior.

This effort is also inspired by the Wheeler dictum that mass and energy are essentially derivatives of information. In the orientable toroidal topology of the universe first considered, the challenge would be to create two sets of entangled qubits representing the exteriors of the wormhole and enabling a sustainable open duct by creating the equivalent of the negative energy pulse in a quantum computer (Figure 5.25). This can be achieved by a rotating magnetic field that induces spin rotation in a ring of qubits representing one of the wormhole exteriors.

Thus, in essence, the wormhole would be created through entanglement of qubits and its sustainability, as qubit passage would be ensured by recreating the cosmic engine fueled by dark energy in the guise of a "pulse of dark energy deficiency" traveling along the wormhole. In this way, dark energy could be recreated in a quantum computer as the enabler and sustainer of the primeval wormhole of the universe.

However, to generate dark fermions, a PNOW with quotient-space defined in Figure 5.17 would need to be implemented in the quantum computer. This would require the antiparallel connection of the outer layers of two quantum-gravity autoencoders as depicted in Figure 5.21b, recreating a qubit passage with 2-dimensional y_5-involved cross sections mapped as a Klein bottle (Figure 5.26).

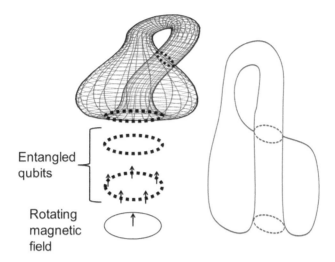

Entangled qubits

Rotating magnetic field

FIGURE 5.26 Scheme of a PNOW recreated in a quantum computer.

5.10 THE PORTAL TO THE DARK UNIVERSE COUPLES DARK ENERGY EXPENDITURE WITH A DYNAMIC EQUILIBRIUM BETWEEN DARK AND DETECTABLE MATTER

Chapter 1 describes the disconcerting but established phenomenology that accounts for the existence of dark matter and dark energy in the standard 4-dimensional space-time. A quintessentially encoded ur-space-time is needed to account for dark matter, as shown in Chapter 4. Yet, dark matter appears likely to be more than one thing. A portion of dark matter stays as ur-particles, created early on in the universe history, while another portion, the dark fermions, are in dynamic equilibrium with detectable matter, as previously shown. The ur-Higgs and its partners, quintessential particles with no geometric dilution into the standard spatial coordinates, account for 81.28% of dark matter. The concentration of ur-particles is ever decreasing in our universe, which undergoes accelerated expansion. On the other hand, the dark fermions in dynamic equilibrium with detectable matter constitute 18.72% of dark matter. Since "matter in all forms," including dark energy, is *at present* 26.7% dark matter and 5% detectable matter (Chapter 1), the dark fermion proportion at 18.72% of dark matter corresponds to 15.77% of all matter (dark + detectable) in the universe, which is exactly equal to the percentage of all matter represented by detectable matter. In turn, the

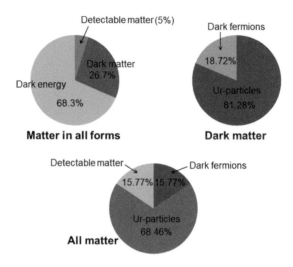

FIGURE 5.27   Estimated composition of "all forms of matter," dark matter, and all matter in the present-day universe.

15.77% of all matter, representing either the proportion of dark fermions or detectable matter, corresponds, *at present*, to 5% of all forms of matter. The proportions in the current estimated composition of the universe are depicted in Figure 5.27.

In this chapter, we have provided a vantage point to identify dark energy as the surplus in vacuum energy taken from the amount required to fuel the cosmic engine that maintains the PNOW in an expanding universe. The concentration of this surplus is indeed constant since dark energy is not subject to dilution associated with vacuum creation. This fact stands in contrast with the concentration of dark matter, including ur-particles and dark fermions, which are in dynamic equilibrium with detectable matter. Clearly, the dark matter concentration decreases as it gets progressively diluted concurrently with vacuum growth over time. The sustainability of the PNOW, depicted schematically in Figures 5.28 and 5.29, ensures the dynamic equilibrium between dark fermions and detectable matter, a contribution that maintains the overall mass of dark matter constant while the overall concentration of dark energy is also kept constant.

To conclude, the dynamic equilibrium between dark matter and detectable matter is maintained by a cosmic engine fueled by vacuum energy that sustains the portal to the dark universe while maintaining a constant concentration of dark energy in an ever-expanding universe.

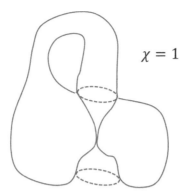

$\chi = 1$

FIGURE 5.28   Schematics of the Klein bottle at unity aspect ratio representing a 2-dimensional cross section for the PNOW.

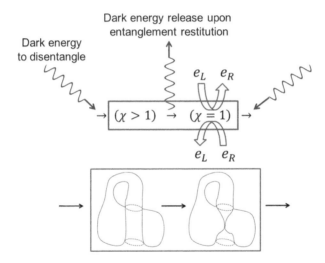

Dark energy release upon entanglement restitution

Dark energy to disentangle

e_L e_R

$(\chi > 1)$ → $(\chi = 1)$ →

e_L e_R

FIGURE 5.29   Schematics of the operation of the cosmic engine fueled by vacuum energy that sustains the portal to the dark universe. The dynamic equilibrium between dark matter and detectable matter is coupled to the production of dark energy (the surplus vacuum energy in each engine cycle) at a constant concentration in an expanding universe.

REFERENCES

1. Maldacena J, Susskind L (2013) Cool horizons for entangled black holes. *Fortsch Phys* 61, 781–811.
2. Bohm D, Aharonov Y (1957) Discussion of Experimental Proof for the Paradox of Einstein, Rosen and Podolsky. *Phys Rev* 108: 1070–1079.
3. Adler R, Casey B, Jacob O (1995) Vacuum catastrophe: An elementary exposition of the cosmological constant problem. *Am J Phys* 63: 620–626.

4. Weinberg S (2008) *Cosmology*. Oxford University Press, New York, USA.
5. Fisher P (2022) *What is Dark Matter?* Princeton University Press, Princeton, NJ.
6. Lee TD, Yang CN (1956) Question of Parity Violation in Weak Interactions. *Phys Rev* 104:254–258.
7. Visser M (1995) *Lorentzian Wormholes: From Einstein To Hawking*. Washington University Press, St. Louis, Missouri.

CHAPTER 6

Physical Footprints of a Computer-Simulated Universe

The Artifactual Nature of the Dark Universe

"With relief, with humiliation, with terror, he understood that he was also an illusion, that someone else was dreaming him."

The Circular Ruins
JORGE LUIS BORGES

In this chapter we critically examine physical footprints that support the conjecture that the universe is simulated in a computer by an advanced civilization capable of cosmic manipulation. The physical evidence supporting the simulation scenario is shown to be hiding in plain sight, in the very laws of contemporary physics. On the other hand, the artifactual nature of the dark universe within the simulation scenario is established and instills confidence in validity of the physical laws that we steadfastly trust.

6.1 THE SIMULATION HYPOTHESIS

As shown in Chapter 5, an assemblage of two replicas of an autoencoder may be deployed to recreate a primeval non-orientable wormhole fueled

DOI: 10.1201/9781003501534-7

by dark energy and purportedly responsible for the dynamic equilibrium between dark and visible matter. Furthermore, following the dictates of statistical mechanics, we can even confer physical reality to the AI system [1], much along the lines of Wheeler's dictum "*it from bit.*" These forays into mathematical cosmology empowered by AI prompt the daring hypothesis that the universe is a simulation, already put forth by the philosopher Nick Bostrom [2].

The Bostrom hypothesis has been assigned a striking 50:50 chance of being true, but the science put forth in Chapter 5 will surely slant this ratio toward a more favorable probability. The hypothesis has been also subject to derision, since it was deemed to be a non-scientific idea, incapable of yielding a falsifiable prediction. This is surely incorrect in light of the cosmological implications of the AI–based metamodel of the universe described in Chapter 5 and the AI–empowered technology for universe reproduction implemented in Chapter 3. There are plenty of physical tests to prod the simulated universe generated by the annular holographic autoencoder, especially probing the two-stroke engine that sustains the primeval wormhole and the quantum physical constants arising thereof. If nothing else, the simulation scenario has passed the ultimate test by providing a provisional solution to the cosmological-constant–vacuum-catastrophe problem.

The idea of a simulated universe, today attributed to a philosopher, finds a precursor in the metaphysical musings of Hinduism, where the deity Vishnu is often depicted as *Anantasayana* (i.e., lying on the serpent *Ananta*) in order to dream the universe into reality (Figure 6.1). Furthermore, in manifold guise, the idea also belonged *mutatis mutandis* to a literary province. Translate the computer science term "simulation" as "dream" and we get the simulation hypothesis in literary embodiments such as the story of the butterfly dreamt by Zhuangzi, the play "La Vida es Sueño" (Life Is a Dream) by Pedro Calderón de la Barca, or the short story "The Circular Ruins" by Jorge Luis Borges, where a man comes to realize that he is being dreamt by another man.

In the scientific realm, the simulation hypothesis has now passed an important test: It has materialized in a physical metamodel of the universe known as holographic autoencoder, an AI system in a physical embodiment capable of reconciling general relativity and quantum mechanics (Chapter 3). This is no minor feat since the respective material scales to which those theories apply are incommensurably different and hence the theories were deemed irreconcilable until the attention was drawn to

FIGURE 6.1 *Vishnu Anantasayana.* Deity Vishnu, lying on the serpent Ananta, dreams the universe into reality. Mural from Dasavatara Temple, Deogarh, Uttar Pradesh, India (Wikimedia Commons, in the public domain).

entangled black holes (Chapter 5). There is at least the certainty that one such object exists in the form of the primeval wormhole, as it is inherent to the universe's compact, boundaryless, and multiply connected topology. Since the simulated universe finds its supreme embodiment in the operation of the dark-energy–fueling engine that sustains the primeval wormhole, we may say that the quantum gravity conundrum has been resolved, at least vis-à-vis the standards laid out in Chapters 3–5.

6.2 IS THE UNIVERSE AMENABLE TO BE SIMULATED?

To confer physical meaning to the simulation hypothesis we need to first decide whether the universe is amenable to be recreated in the conventional sense that we ascribe to the phrase "simulation in a computer." Simulating a process at a specific level of coarse graining, anywhere from organs to quarks, assumes that the process can be thoroughly captured in a reality cell with periodic boundary conditions and that such periodic boundary conditions are physically meaningful in the sense that they do not introduce artifacts affecting the outcome of the simulated process.

If the universe were infinite, extending this line of thought to a simulation of the whole universe may become daunting because we do not really know whether large-scale events – such as patterns of temperature distribution in the cosmic microwave background – may become artificially

Simulation $\mathbb{R}^3/\mathbb{Z}^3$

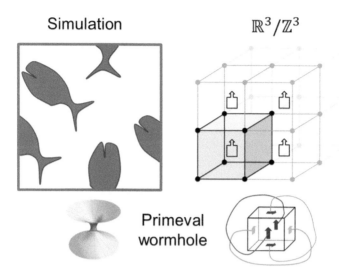

Primeval
wormhole

FIGURE 6.2   The spatial cross section of the universe as the quotient space $\mathbb{R}^3/\mathbb{Z}^3$ equivalent to a finite (however enormous) reality cell with periodic boundary conditions. Thus, in contrast with an infinite universe, the actual topology of the universe makes it amenable to be generated in a computer simulation. The simulation must recreate the primeval wormhole inherent to the toroidal manifold $\mathbb{R}^3/\mathbb{Z}^3$, which in turn requires that we postulate the existence of the dark universe.

truncated. On the other hand, we are now persuaded that the spatial manifold of the universe is not infinite ($\mathbb{R}^3$) but finite; specifically, the multiply connected compact manifold $\mathbb{R}^3/\mathbb{Z}^3$ that necessarily spans a primeval wormhole and is compatible with the big bang birth scenario, as described in Chapter 5. In other words, the spatial manifold is actually the quotient space $\mathbb{R}^3/\mathbb{Z}^3$ equivalent to a reality cell with periodic boundary conditions (Figure 6.2). Thus, in contrast with an infinite universe, *the actual topology of the universe makes it amenable to be generated in a computer simulation.* Furthermore, again in contrast with an infinite universe, its compactness allows for the expansion of the simulated reality cell, a process assumed to be fueled by dark energy.

6.3 THE DARK UNIVERSE AS A COMPUTATIONAL ARTIFACT IN THE SIMULATION OF THE UNIVERSE

As discussed in the previous chapters, the physical embodiment of the holographic autoencoder bestows materiality to the $\mathbb{R}^3/\mathbb{Z}^3$ (3-torus) spatial cross section of the universe. In so far as we judiciously assume that

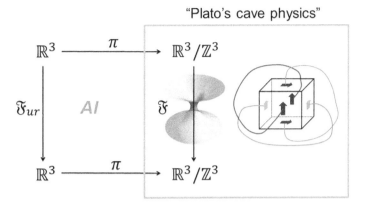

FIGURE 6.3 Commutative scheme representing the AI–empowered autoencoder leveraged to simulate the universe. The dark universe becomes an artifact of the projection onto the quotient space ($\mathbb{R}^3 \to \mathbb{R}^3/\mathbb{Z}^3$) within the commutative scheme that enables the universe simulation by lifting the primeval wormhole inherent in $\mathbb{R}^3/\mathbb{Z}^3$ to the level of its "featureless" universal covering.

this manifold represents the actual physical space, its inherent primeval wormhole is sustained by dark energy as informed by the AI–empowered simulation (Chapter 5). On the other hand, the autoencoder system simulates the universe based on the commutative scheme presented in Figure 6.3 whereupon, at the level of the universal covering $\mathbb{R}^3$ with projection $\pi : \mathbb{R}^3 \to \mathbb{R}^3/\mathbb{Z}^3$ [3], there is no primeval wormhole and, consequently, no dark universe to account for. In other words, *the dark universe is an artifact of the projection onto the quotient space ($\mathbb{R}^3 \to \mathbb{R}^3/\mathbb{Z}^3$) within a commutative scheme that enables the universe simulation by lifting the primeval wormhole inherent in $\mathbb{R}^3/\mathbb{Z}^3$ to the level of its "featureless" universal covering. "Plato's cave" physical account given by the flow map $\mathfrak{F}$ on $\mathbb{R}^3/\mathbb{Z}^3$ contains a primeval singularity that has no reality in the lifted flow map $\mathfrak{F}_{ur}$ of the ur-universe.*

Underlying this discussion is the AI–empowered scheme wherein the simulated universe is defined by Plato's cave flow map $\mathfrak{F} : \mathbb{R}^3 / \mathbb{Z}^3 \to \mathbb{R}^3 / \mathbb{Z}^3$, while the ur-universe is operationally represented by leveraging the surjective map $\pi : \mathbb{R}^3 \to \mathbb{R}^3/\mathbb{Z}^3$ within the commutative diagram that determines the relation: $\pi \circ \mathfrak{F}_{ur} = \mathfrak{F} \circ \pi$, as shown in Figure 6.3.

6.4 PHYSICAL FOOTPRINTS OF A SIMULATED UNIVERSE

Any computer simulation has embossed in it the imprint of the processor responsible for its generation. In our case, this imprint should surface in the discoverable universe as an "anomaly" or, rather, in a physical law that defies "intuition." However, our awareness of this anomaly is numbed, clouded by the extreme familiarity with the universe acquired through our conscience, a familiarity sculpted by the accretion of all our experiences that begin at the cradle.

Quantum physics offers a plethora of perplexities that can be readily accounted for if we assume that we live in a simulated universe. When fleshed out in the setting of the universe as a simulation, the fundamental postulate that quantum phenomena do not materialize until they become registered phenomena becomes far more intellectually digestible. *In the simulation scenario, a huge amount of processor time is saved if the observer or detector plays the ontological role assigned by quantum physics*: There is simply no obvious need to materialize phenomena when no act of observation or detection is also simulated to trigger the collapse of the wave function concurrently with state disambiguation. Likewise, quantum entanglement, an object of derision by none other than Albert Einstein, finds a straightforward explanation in the simulation scenario, since locality is immaterial in a virtual reality.

Evidently, if a civilization is simulating our universe having previously constructed some version of the holographic autoencoder, it must be inconceivably more advanced than ours, if nothing else because it has shown to be capable of cosmic manipulation (so far we are not). Yet, regardless of the degree of development, the speed of the processor this civilization was able to deploy cannot conceivably be infinite, an assertion that can be made with certainty. We know with astounding accuracy what the processor speed is!: It is related to the storage capacity of the information required for the processor to perform a single operation. If the category "physical space" is being simulated and the processor performs, say, a single operation per second, then the amount of space encoded as bits of information generated must be translatable into actual physical space spanning ~300,000 km along a single dimension. That is, the processor speed begets the speed of light (~300,000 km/s) in its simulated universe, making it a staggering performer, indeed!

If indeed the speed of light, an absolute upper limit and a key parameter defining our universe, contains the imprint of the processor that simulates it, then this speed must be constant irrespective of the location or trajectory of the observer that measures it. This is because the speed of light is not "physical" but an *object d'art* of the simulation. And this is indeed the case! In fact, this is a major tenet of relativity, and Einstein had to postu-late the slowing down of time for an observer traveling at speeds close to the speed of light (relativistic speeds) in order to accommodate the artifact of constant measured speed of light: For that particular "relativistic" observer (A), the distance travelled by light is shorter than that for an observer (B) that moves at slower (nonrelativistic) speeds, but time runs more slowly for A than for B, so that the speed of light measured by both observers is the same.

In other words, Einstein had to tinker with time to turn the computa-tional artifact accurately described by the proposition "light speed is con-stant" into a "physical" proposition. In this sense, space-time becomes Einstein's way to circumvent, or rather, come to grips with the perceived anomaly introduced by the fact that our universe is a simulation, some-thing he could have not anticipated at the time when his theory of relativ-ity was conceived.

If the universe is indeed a simulation, the processor performance for maximum quark-level resolution (~1 attometer) must be a staggering $f \sim 3 \times \frac{10^8}{10^{-18}}$ Hz $= 3 \times 10^{14}$ Teraflops, which is about 10^{13} times higher than the number of floating-point operations per second accessible to quantum computation. However, there are ways to circumvent the prohibitively fast performance required. This entails using autoencoders trained to distill (encode) the coarse-grained version of reality at human eye resolution, that is, at 10^{-5} m, and decode the information all the way back to a quark-resolved reality in special cases when there is a need for such resolution (i.e., in a lab experiment broadly defined). In a quotient space resolving reality at a "human scale," the required processor speed would be $f \sim 3 \times \frac{10^8}{10^{-5}}$ Hz $= 30$ Teraflops, which is commensurate with the perfor-mance of today's quantum computers. The universe can thus be simulated by two coupled autoencoders fulfilling the commutativity condition for the diagram shown in Figure 6.4.

Quantum mechanical phenomena also find a natural correlate in the simulation scenario. Illustrative examples that merit to be mentioned, and would require further refinement of the scheme presented in Figure 6.4, include the need for an observer to disambiguate the state of the quantum system and the nonlocality of quantum physics due to ubiquitous entan-glement in the quintessential universe W, the ur-universe described at the opening of Chapter 1.

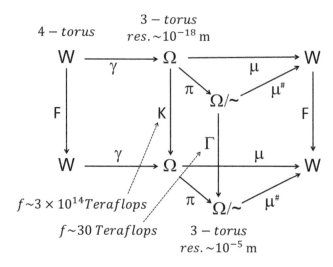

FIGURE 6.4 Commutative diagram representing an AI architecture to simulate the universe with two coupled autoencoders circumventing the prohibitively fast performance required at quark-level resolution. The autoencoders are coupled to distill (encode) the coarse-grained version of reality at human eye resolution (10^{-5} m) and decode this information back to quark-level resolution of reality in special cases when required by an observer/experimenter.

The time slowdown or dilation experienced by the observer that approaches the speed of light is an essential relativistic outcome that finds its natural correlate in the simulated universe scenario: The overloading of the processor that simulates relativistic reality slows down its performance by increasing loading times up to a freeze that occurs as the speed of light is reached by the simulated observer. In practice, in simulations performed on the annular assemblage of holographic autoencoders (Figures 5.10 and 5.21b), it is observed that processor overloading leads to a complete freeze as $\chi \to 1$. In fact, no current AI–based cosmological technology enables us to study the primeval wormhole for levels of disentanglement beyond $E = 18$ (Figure 5.11), corresponding to $\chi \approx 1 + 10^{-32}$.

The topology of the universe was suspected to be of no tangible significance to physics up to this juncture in the history of science. As AI–empowered metamodels make their forays in the field of cosmology, the situation is likely to change rapidly and the possibility that the universe may be a simulation does not appear so farfetched, at least to the open mind.

The physical footprints of a simulation scenario have been shown in this section to be hiding in plain sight, in the very laws of contemporary physics, while the artifactual nature of the dark universe within the simulation scenario should instill confidence in validity of those same physical laws that we unwaveringly uphold.

6.5 AI ARCHITECTURES TO ENCODE THE STANDARD MODEL: PIXELATION OF DEEP REALITY

As noted American physicist John A. Wheeler would have it [4], the physics world is information, while mass and energy are incidentals. This view is enshrined in his aphorism "*it from bit*." Since virtual reality is based on information processing, Wheeler's picture substantiates Bostrom's simulation hypothesis [2], an assertion that becomes especially compelling if we construe the uncertainty principle as a glimpse into deep reality. Thus, the Compton wavelength, $\lambda_C = \dfrac{h}{mc}$, in the photon scattering by a particle of mass m could be interpreted as the pixelation of deep reality. This is so because λ_C represents the spatial uncertainty in the observation of the particle.

The compact, multiply connected nature of the latent manifold for Einstein's space-time makes the universe amenable to be simulated, as shown in Section 6.2. On the other hand, physical footprints of the simulated universe are surely apparent in the quantum physics description of the pixelation of deep reality and in the relativistic parametrization of the processor frequency, as indicated in Section 6.4.

So, if quanta represent pixelation, and virtual reality is the outcome of information processing, and the processor itself has a relativistic footprint, what sort of AI system is actually simulating the universe? The AI models for the dark universe described in Chapters 4 and 5 suggest that this question becomes tantamount to ask: What sort of autoencoder architecture can generate the Standard Model (SM) of particle physics with pixelation tailored to quantum field theory?

Encoding the 15 massive particle fields in the SM would require 15 autoencoders compatible with each other and capable of encoding also photons and gluons regarded as endowers of reality through materialization of observation. The processor for each autoencoder is endowed with performance frequency *f*, associated with the respective Compton length as a measure of pixelation according to the relation:

$$f = \frac{c}{\lambda_C} = mc^2/h. \tag{6.1}$$

This implies that the simulation of the universe at the highest resolution requires using an autoencoder capable of encoding the top quark field (heaviest particle) with $m = 173 \text{GeV}/c^2 \approx 3.08 \times 10^{-25}$ kg and $f \approx 4.19 \times 10^{25}$ Hz $= 4.19 \times 10^{13}$ Teraflops. We name this autoencoder AE. At this colossal level of computer performance, an advanced civilization is capable of encoding all fields in the SM, thereby simulating the entire universe with a single processor.

To save computer power, AE needs to be entrained by other autoencoders, AE_1, AE_2, ..., A_{14}, capable of simulating the universe at coarser levels of resolution. In particular, AE_1, the autoencoder of the Higgs boson ($m = m_H \approx 2.22 \times 10^{-25}$ kg; $f \approx 3.02 \times 10^{13}$ Teraflops) and all the lighter particles ($m < m_H$), should be optimized to be completely compatible with AE, as per the commutativity of diagram in Figure 6.5. Simulations at progressively coarser levels of resolution, carried out by AE_2 (Z-boson), AE_3 (W-boson), ..., AE_{14} (electron neutrino), should be made compatible

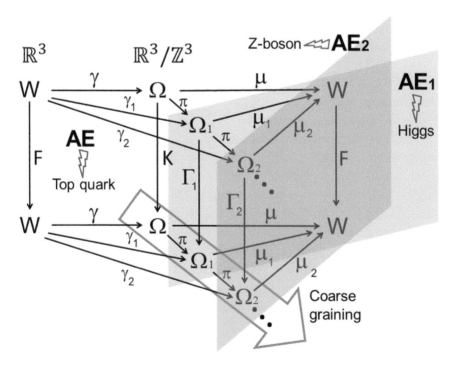

FIGURE 6.5 Commutative diagram representing an AI architecture consisting of 15 coupled (pairwise compatible) autoencoders generating the Standard Model of particle physics at different levels of field resolution. Pixelation of deep space becomes commensurate with the Compton lengths of the respective particles.

with AE and AE_1 in the sense that the diagram (Figure 6.5) describing map compositions is commutative. The entrainment of all autoencoders through the commutativity with AE_1, itself trained at the highest cost by AE, endows the respective particles with mass, in accord with the Higgs mechanism. Once all AEs are trained and optimized fulfilling diagram commutativity (Figure 6.5), the universe may be readily simulated at the coarsest level by AE_{14}, which operates at the lowest cost and encodes only the electron neutrino (ν_e), the lightest massive particle ($m < 0.8$ eV/$c^2 \approx 1.42 \times 10^{-36}$ kg). Thus, the processor performance frequency to simulate the universe at the lowest admissible resolution is estimated at 193 Teraflops. This level of performance seems very much within reach in the foreseeable future.

REFERENCES

1. Fernández A (2022) *Topological Dynamics for Metamodel Discovery with Artificial Intelligence*. Chapman & Hall/CRC, Taylor and Francis, London, UK.
2. Bostrom N (2003) Are We Living in a Computer Simulation? *Philos Quaterly* 53: 243–255.
3. Fernández A, Sinanoglu O (1982) The Lifting of an Inonu-Wigner Contraction at the Level of Universal Coverings. *J Math Phys* 23: 2234–2235.
4. Wheeler JA (1982) The Computer and the Universe. *Int J Theor Phys* 21: 557–572.

Epilogue

Gauge Symmetry of Dark Matter Associated with the Conservation of Geometric Dilution Leads to Dark Boson Prediction

> *"The reasonable man adapts himself to the world,*
> *the unreasonable one persists in trying to adapt*
> *the world to himself. Therefore all progress*
> *depends on the unreasonable man."*
> GEORGE BERNARD SHAW

As emphasized throughout the book, the Standard Model (SM) constitutes a towering achievement in particle physics. Its predictive value is simply staggering, making it the benchmark of quantum field theory. A cornerstone for its success is given by the way it introduces continuous internal symmetries for elementary particles that are in turn represented as excitations of their respective fields. In accordance with Noether's theorem, the SM introduces an internal symmetry in a particle field in such a way that the invariance of the field-defining Lagrangian under a local symmetry transformation becomes the determinant of the conservation of a physical quantity attributed to the particle, such as charge, spin, and so on.

The local symmetry invariance requires that the kinetic energy be defined in terms of covariant derivatives and that potential energy terms be added, so that the effective Lagrangian includes terms defining a "gauge field" that bestows the invariance to the particle field. In essence, we may

DOI: 10.1201/9781003501534-8

say that the SM is a gauge theory because the conservation of physical quantities is achieved through the incorporation of gauge fields that ensure local internal symmetries of the Lagrangian. Local excitations of gauge fields represent gauge bosons, therefore conservation of physical quantities in fermions undergoing physical transformations are achieved through interaction with gauge bosons. This interaction is interpreted in the SM by assuming that the bosons communicate the force associated with the gauge field that in turn determines the internal symmetry of the fermion field.

Consequently, in the SM, fermions cannot be considered isolated entities because their fields incorporate gauge terms reflective of interactions with gauge bosons. For example, the field of the electron is endowed with an internal symmetry associated with charge conservation if and only if the fermion Lagrangian incorporates terms determining the gauge field needed to generate the local internal symmetry under Noether's theorem. This gauge field has been established to correspond to the photon; therefore, the SM cannot represent the electron in isolation but computes its kinetic energy by squaring the covariant derivate along the photon internal coordinate and includes gauge terms in the electron's potential energy. Thus, according to quantum field theory, the charge invariance of the electron can be upheld only by regarding the electron in conjunction with the photon.

This representation of fermionic fields in conjunction with gauge fields finds its equivalent in the autoencoders for the SM, described in Chapters 2 and 6. Thus, just like in the local symmetry invariance of particle Lagrangians, no autoencoder can operate in isolation: They are optimized for full commutativity with one another [1], so the local internal symmetry in the autoencoder for a fermion with a particular Compton length is bestowed via the entrainment of that autoencoder by another autoencoder that encompasses the required gauge field. For example, a down quark may be encoded at far lower processor frequency than that required for the W-boson but high enough to also encode the electron and the electron antineutrino. This implies that the electroweak force communicated in the process of beta decay (Figure 2.13), whereby the down quark becomes an up quark, with concomitant release of an electron and electron antineutrino, cannot materialize in the autoencoder for the down quark (AE_d) operating in isolation. For the process to materialize, it is required that the AE_d be entrained by another autoencoder, AE_W, with high enough processor frequency to encode the W-boson (Figure E1). This entrainment is

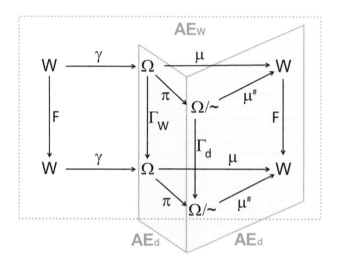

FIGURE E1 Commutative diagram materializing the entrainment of the optimized autoencoder for the down quark (AE_d) by the optimized autoencoder for the W-boson (AE_W), which has a higher frequency processor ($f_W > f_d$). The entrainment incorporates the gauge field for the down quark, endowing it with local internal symmetry and enabling the beta-decay process that converts the down quark into an up quark via interaction with the gauge boson (Figure 2.13). The equivalence relation "~" indicates classes modulo the longer wavelength (c/f_d), so that the latent quotient space ($\Omega/{\sim}$, π) represents the coarse graining associated with the wavelength change $c/f_W \to c/f_d$.

achieved by parametrically optimizing both autoencoders so the diagram relating them is commutative (Figure E1). This argument shows that AI is capable of constructing the gauge fields through entrainment of particle autoencoders by "gauge autoencoders" in such a way that both autoencoders are coupled through diagram commutativity. In this way, internal local symmetries are encoded in the AI model.

On the other hand, according to the AI model of the dark universe, all forms of matter including dark matter have a conserved quantity that is not featured in the SM: The geometric dilution of the particle field into the "observable spatial dimensions": $v = -\log\cos\alpha$, where $tg\alpha$ is the quotient between the projection onto observable dimensions over the projection onto dark dimensions (Figure E2). Thus, for dark matter we get $v = 0$ and the invariance of this physical quantity prompts us to enquire about the hitherto unknown internal symmetry of dark matter. This is the first question that needs to be addressed as we implement a gauge theory for the dark universe.

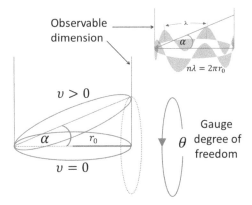

FIGURE E2 Newly discovered gauge symmetry associated with the conservation law described by v = constant. The Lagrangian for geometrically undiluted "dark" particles (v = 0) is invariant under local θ-symmetry transformations, while particles with $v > 0$ require interaction with a gauge boson to guarantee invariance under local θ-symmetry transformations. The prediction of the "dark boson" follows from the discovery of the new conservation law presented in this book.

Any extension of the SM to encompass the dark dimension would be required to introduce a gauge symmetry associated with the conservation of v. Such gauge symmetry, as yet unknown, is expected to be discoverable since it is enshrined in the AI model for the dark universe, as it follows from the previous discussion. After all, the AI model enabled the discovery of the conservation law involving the constancy of v.

In the case of dark matter, with its geometrically undiluted kinetic energy, the continuous symmetry associated with the conservation law $v = 0$ is the rotation perpendicular to the circular dark dimension corresponding to an angular degree of freedom θ (Figure E2). The invariance of the Lagrangian under the local symmetry is ensured because the projection of the particle field along all dimensions – observable and dark – remains invariant under the generic transformation $\theta \to \theta + \theta_0$, $\forall\ \theta_0$. However, for a particle with geometric dilution $v = -\log \cos \alpha > 0$, there is no invariance of the field projections – and therefore of the Lagrangian – under local symmetry transformations (Figure E2). *This implies that the conservation of geometric dilution of the particle field requires a new gauge field to ensure invariance of the Lagrangian under local θ-symmetry. Thus, we are predicting the existence of a new gauge boson and hereby name it the dark boson.* In fact, there should be three dark bosons since the visible dimension combining with the dark dimension to yield geometric dilution can be permuted. On the other hand, only one dilution parameter can be specified for each permutation since the rotation operators associated with visible axes do not commute. Thus, a gauge theory for the dark sector should be a SU(2) theory, or rather a SU(2)×U(1) theory, when including the internal rotational symmetry for the dark circular dimension.

A clear delineation of the experimental setting required to corroborate the prediction of the dark bosons based on the newly discovered gauge symmetry will surely motivate research efforts for years to come.

REFERENCE

1. Fernández A (2022) *Topological Dynamics for Metamodel Discovery with Artificial Intelligence*. Chapman & Hall/CRC, Taylor and Francis, London, UK.

Index